Yearning for the Impossible

Impossible

The Surprising Truths
of Mathematics

Second Edition

Yearning for the Impossible

The Surprising Truths of Mathematics

Second Edition

John Stillwell

CRC Press

Taylor & Francis Group

Boca Raton London New York

CRC Press is an imprint of the
Taylor & Francis Group, an **informa** business

A CHAPMAN & HALL BOOK

CRC Press
Taylor & Francis Group
6000 Broken Sound Parkway NW, Suite 300
Boca Raton, FL 33487-2742

Printed on acid-free paper
Version Date: 20180404

International Standard Book Number-13: 978-1-1385-9621-4 (Hardback)
International Standard Book Number-13: 978-1-1385-8610-9 (Paperback)

Visit the Taylor & Francis Web site at
http://www.taylorandfrancis.com

and the CRC Press Web site at
http://www.crcpress.com

To Elaine

Preface to the Second Edition

For the past 12 years I have taught a course based on this book at the University of San Francisco. It is a first-year seminar, intended mainly for students in non-mathematical courses. In keeping with the theme of the book, the seminar is entitled "Mathematics and the Impossible," and it has been pitched as follows:

> This course is a novel introduction to mathematics and its history. It puts the difficulties of the subject upfront by enthusiastically tackling the most important ones: the seemingly impossible concepts of irrational and imaginary numbers, the fourth dimension, curved space, and infinity. Similar "impossibilities" arise in music, art, literature, philosophy, and physics—as we will see—but math has the precision to separate actual impossibilities from those that are merely apparent. In fact, "impossibility" has always been a spur to the creativity of mathematicians, and a major influence on the development of math. By focusing reason and imagination on several apparent impossibilities, the course aims to show interesting math to students whose major may be in another field, and to widen the horizons of math students whose other courses are necessarily rather narrowly focused.

Thus the aim of the seminar is to introduce students to the ideas of mathematics, rather than to drill them in mathematical techniques. But (as all mathematicians know) mathematics is not a spectator sport. So it has been my practice to give exercises alongside classroom discussion of the book, to ensure that students grapple with interesting ideas for themselves. Since other teachers may also decide to give courses based on the book, I have decided to produce this second edition, which includes the exercises I have used with my class and many more.

The exercises are distributed in small batches after most sections, and a batch should be attempted by students after the corresponding section has been discussed in class. Some of the exercises are quite routine—intended to reinforce the ideas just discussed—but other exercises serve to extend the ideas and develop them in interesting directions. There are also exercises that fill gaps by supplying proofs of results claimed without proof in the text, or by answering questions that are likely to arise. The more advanced exercises are accompanied by commentary that explains their context and background. In this way I have been able to cover several topics that will be of interest to more ambitious readers.

Of course, it is by no means necessary to use the book in the context of a course. I hope that it will continue to be used for recreational reading or self-study, and that even the casual reader will be tempted to try some of the exercises. However, for those teachers who have wished to give a course based on the book, I hope that teaching from it has now become much easier.

John Stillwell
San Francisco

Preface

The germ of this book was an article I called (somewhat tongue-in-cheek) "Mathematics Accepts the Impossible." I wrote it for the Monash University magazine *Function* in 1984 and its main aim was to show that the "impossible" figure shown above (the Penrose tribar) is actually not impossible. The tribar exists in a perfectly reasonable space, different from the one we think we live in, but nevertheless meaningful and known to mathematicians. With this example, I hoped to show a general audience that mathematics is a discipline that demands imagination, perhaps even fantasy.

There are many instances of apparent impossibilities that are important to mathematics, and other mathematicians have been struck by this phenomenon. For example, Philip Davis wrote in *The Mathematics of Matrices* of 1965:

> It is paradoxical that while mathematics has the reputation of being the one subject that brooks no contradictions, in reality it has a long history of successfully living with contradictions. This is best seen in the extensions of the notion of number that have been made over a period of 2500 years ... each extension, in its way, overcame a contradictory set of demands.

Mathematical language is littered with pejorative and mystical terms such as irrational, imaginary, surd, transcendental that were once used to ridicule supposedly impossible objects. And these are just terms applied to numbers. Geometry also has many concepts that seem impossible to most people, such as the fourth dimension, finite universes, and curved space—yet geometers (and physicists) cannot do without them. Thus there is no doubt that mathematics flirts with the impossible, and seems to make progress by doing so.

The question is: why?

I believe that the reason was best expressed by the Russian mathematician A. N. Kolmogorov in 1943 [31, p. 50]:

> At any given moment there is only a fine layer between the "trivial" and the impossible. Mathematical discoveries are made in this layer.

To put this another way: mathematics is a story of close encounters with the impossible because *all its great discoveries are close to the impossible*. The aim of this book is to tell the story, briefly and with few prerequisites, by presenting some representative encounters across the breadth of mathematics. With this approach I also hope to capture some of the feeling of *ideas in flux*, which is usually lost when discoveries are written up. Textbooks and research papers omit encounters with the impossible, and introduce new ideas without mentioning the confusion they were intended to clear up. This cuts long stories short, but we have to experience some of the confusion to see the need for new and strange ideas.

It helps to know why new ideas are needed, yet *there is still no royal road to mathematics*. Readers with a good mathematical background from high school should be able to appreciate all, and understand most, of the ideas in this book. But many of the ideas are hard and there is no way to soften them. You may have to read some passages several times, or reread earlier parts of the book. If you find the ideas attractive you can pursue them further by reading some of the suggested literature. (This applies to mathematicians too, some of whom may be reading this book to learn about fields outside their specialty.)

As a specific followup, I suggest my book *Mathematics and Its History*, which develops ideas of this book in more detail, and reinforces them with exercises. It also offers a pathway into the classics of mathematics, where you can experience "yearnings for the impossible" at first hand.

Several people have helped me write, and rewrite, this book. My wife, Elaine, as usual, was in the front line; reading several drafts and making the first round of corrections and criticisms. The book was also read carefully by Laurens Gunnarsen, David Ireland, James McCoy, and

Abe Shenitzer, who gave crucial suggestions that helped me clarify my general perspective.

Acknowledgments. I am grateful to the M. C. Escher Company-Baarn-Holland for permission to reproduce the Escher works *Waterfall*, shown in Figure 8.1, *Circle Limit IV*, shown in Figure 5.26, and the two transformations of *Circle Limit IV* shown in Figures 5.26 and 5.27. The Escher works are copyright (2005) The M. C. Escher Company – Holland. All rights reserved. www.mcescher.com.

I also thank the Artists Rights Society of New York for permission to reproduce the Magritte picture *La reproduction interdite* shown in Figure 8.9. This picture is copyright (2006) C. Herscovici, Brussels/Artists Rights Society (ARS), New York.

John Stillwell
South Melbourne, February 2005
San Francisco, December 2005

Contents

Chapter 1

The Irrational

Preview

What are numbers and what are they for? The simplest answer is that they are the *whole numbers* 1, 2, 3, 4, 5, ... (also called the *natural numbers*) and that they are used for counting. Whole numbers can also be used for *measuring* quantities such as length by choosing a unit of measurement (such as the inch or the millimeter) and counting how many units are in a given quantity.

Two lengths can be accurately compared if there is a unit that measures them both exactly—a *common measure*. Figure 1.1 shows an example, where a unit has been found so that one line is 5 units long and the other is 7 units long. We can then say that the lengths are in the *ratio* of 5:7.

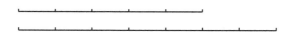

Figure 1.1: Finding the ratio of lengths.

If a common measure exists for any two lines, then any two lengths are in a natural number ratio. *Mathematicians once dreamed of such a world—in fact, a world so simple that natural numbers explain everything. However, this "rational" world is impossible.*

The ancient Greeks discovered that there is no common measure for the side and diagonal of a square. We know that when the side is 1

1

the diagonal is $\sqrt{2}$, hence $\sqrt{2}$ is *not* a ratio of natural numbers. For this reason, $\sqrt{2}$ is called *irrational*.

Thus $\sqrt{2}$ lies outside the rational world, but is it nevertheless possible to treat irrational quantities as numbers?

1.1   The Pythagorean Dream

> It is clear that two scientific methods will lay hold of and deal with the whole investigation of quantity; arithmetic, absolute quantity, and music, relative quantity.
>
> Nicomachus, *Arithmetic*, Chapter III

In ancient times, higher learning was divided into seven disciplines. The first three—grammar, logic, rhetoric—were considered easier and made up what was called the *trivium* (which is where our word "trivial" comes from). The remaining four—arithmetic, music, geometry, astronomy—made up the advanced portion, called the *quadrivium*. The four disciplines of the quadrivium are naturally grouped into two pairs: arithmetic and music, and geometry and astronomy. The connection between geometry and astronomy is clear enough, but how did arithmetic become linked with music?

According to legend, this began with Pythagoras and his immediate followers, the Pythagoreans. It comes down to us in the writings of later followers such as Nicomachus, whose *Arithmetic* quoted above was written around 100 CE.

Music was linked to arithmetic by the discovery that harmonies between the notes of plucked strings occur when the lengths of the strings are in small whole number ratios (given that the strings are of the same material and have the same tension). The most harmonious interval between notes, the *octave*, occurs when the ratio of the lengths is 2:1. The next most harmonious, the *fifth*, occurs when the ratio of the lengths is 3:2, and after that the *fourth*, when the ratio is 4:3. Thus musical intervals are "relative" quantities because they depend, not on actual lengths, but on the ratios between them. Seeing numbers in music was a revelation to the Pythagoreans. They thought it was a glimpse of something greater: the all-pervasiveness of number and harmony in the universe. In short, *all is number*.

We now know that there is a lot of truth in this Pythagorean dream, though the truth involves mathematical ideas far beyond the natural numbers. Still, it is interesting to pursue the story of natural numbers in music a little further, as later developments clarify and enhance their role.

The octave interval is so harmonious that we perceive the upper note in some way as the "same" as the lower. And we customarily divide the interval between the two into an eight-note scale (hence the terms "octave," "fifth," and "fourth")—do, re, mi, fa, so, la, ti, do—whose last note is named the same as the first so as to begin the scale for the next octave.

But why do notes an octave apart sound the "same"? An explanation comes from the relationship between the length of a stretched string and its *frequency of vibration*. Frequency is what we actually hear, because notes produced by (say) a flute and a guitar will have the same pitch provided only that they cause our eardrum to vibrate with the same frequency. Now if we halve the length of a string it vibrates twice as fast, and more generally if we divide the length of the string by n, its frequency is multiplied by n. This law was first formulated by the Dutch scientist Isaac Beeckman in 1615. When combined with knowledge of the way a string produces a tone (actually consisting of many notes, which come from the *modes of vibration* shown in Figure 1.2), it shows that each tone *contains* the tone an octave higher. Thus it is no wonder that the two sound very much the same.

A string has infinitely many simple modes of vibration: the fundamental mode in which only the endpoints remain fixed, and higher modes in which the string vibrates as if divided into $2, 3, 4, 5, \ldots$ equal parts. If the fundamental frequency is f, then the higher modes have frequencies $2f, 3f, 4f, 5f, \ldots$ by Beeckman's law.

When the string is plucked, it vibrates in all modes simultaneously, so in theory all these frequencies can be heard (though with decreasing volume as the frequency increases, and subject to the limitation that the human ear cannot detect frequencies above about 20,000 vibrations per second). A string with half the length has fundamental frequency $2f$—an octave higher—and higher modes with frequencies $4f, 6f, 8f, 10f, \ldots$. Thus all the frequencies of the half-length string are among those of the full-length string.

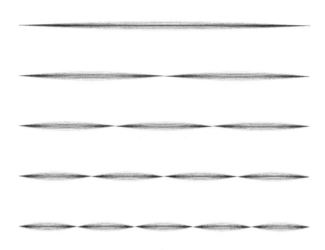

Figure 1.2: Modes of vibration.

Since frequency doubling produces a tone that is "the same only higher," repeated doubling produces tones that are perceived to *increase in equal steps*. This was the first observation of another remarkable phenomenon: *multiplication perceived as addition*. This property of perception is known in psychology as the *Weber-Fechner law*. It also applies, approximately, to the perception of volume of sound and intensity of light. But for pitch the perception is peculiarly exact and it has the octave as a natural unit of length.

The Pythagoreans knew that addition of pitch corresponds to multiplication of ratios (from their viewpoint, ratios of lengths). For example, they knew that a fifth (multiplication of frequency by 3/2) "plus" a fourth (multiplication of frequency by 4/3) equals an octave because

$$\frac{3}{2} \times \frac{4}{3} = 2.$$

Thus the fifth and fourth are natural steps, smaller than the octave. Where do the other steps of the eight-note scale come from? By adding more fifths, the Pythagoreans thought, but in doing so they also found some limitations in the world of natural number ratios.

If we add two fifths, we multiply the frequency twice by 3/2. Since

$$\frac{3}{2} \times \frac{3}{2} = \frac{9}{4},$$

the frequency is multiplied by 9/4, which is a little greater than 2. Thus the pitch is raised by a little over an octave. To find the size of the step over the octave we divide by 2, obtaining 9/8. The interval in pitch corresponding to multiplication of frequency by 9/8 corresponds to the second note of the scale, so it is called a *second*. The other notes are found similarly: we "add fifths" by multiplying factors of 3/2 together, and "subtract octaves" by dividing by 2 until the "difference" is an interval less than an octave (that is, a frequency ratio between 1 and 2).

After 12 fifths have been added, the result is very close to 7 octaves, and one also has enough intervals to form an eight-note scale, so it would be nice to stop. The trouble is, 12 fifths are not *exactly* the same as 7 octaves. The interval between them corresponds to the frequency ratio

$$\left(\frac{3}{2}\right)^{12} \div 2^7 = \frac{3^{12}}{2^{19}} = \frac{531441}{524288} = 1.0136....$$

This is a very small interval, called the *Pythagorean comma*. It is about 1/4 of the smallest step in the scale, so one fears that the scale is not exactly right. Moreover, the problem cannot be fixed by adding a larger number of fifths. A sum of fifths is *never* exactly equal to a sum of octaves. Can you see why? The explanation is given in the last section of this chapter.

This situation seems to threaten the dream of a "rational" world, a world governed by ratios of natural numbers. However, we do not know whether the Pythagoreans noticed this threat in the heart of their favorite creation, the arithmetical theory of music. What we do know is that the threat became clear to them when they looked at the world of geometry.

Exercises

The interval from the first note of the scale to the second is also called a *tone*, and the seven intervals in the usual scale C, D, E, F, G, A, B, C (from C to D, from D to E, ..., from B to C) are tone, tone, semitone, tone, tone, tone, semitone

respectively. Thus the interval from C to the next C is six tones, which should correspond to the frequency ratio of 2.

1.1.1 If a tone corresponds to a frequency ratio of 9/8, as the Pythagoreans thought, explain why an interval of six Pythagorean tones corresponds to a frequency ratio of $9^6/8^6$.

1.1.2 Show that $9^6/8^6$ is *not* equal to 2.

1.1.3 Show that, in fact, $9^6/8^6$ divided by 2 is $\frac{531441}{524288}$ (a "Pythagorean comma").

In music today the interval from C to the C one octave higher is divided into 12 equal semitones with the help of extra notes called $C^{\#}$ (lying between C and D, and pronounced "C sharp"), $D^{\#}$, $F^{\#}$, $G^{\#}$, and $A^{\#}$. These are the black keys on the piano.

1.1.4 Which note divides the octave from C to C into two equal intervals?

1.1.5 Find notes which divide the octave from C to C into

 a. Three equal intervals.

 b. Four equal intervals.

 c. Six equal intervals.

1.2   The Pythagorean Theorem

The role of natural numbers in music may be the exclusive discovery of the Pythagoreans, but the equally remarkable role of natural numbers in geometry was discovered in many other places—Babylonia, Egypt, China, India—in some cases before the Pythagoreans noticed it. As everyone knows, the *Pythagorean theorem* about right-angled triangles states that the square on the hypotenuse c equals the sum of the squares on the other two sides a and b (Figure 1.3).

The word "square" denotes the *area* of the square of the side in question. If the side of the square has length l units, then its area is

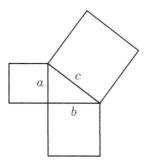

Figure 1.3: The Pythagorean theorem.

naturally divided into $l \times l = l^2$ unit squares, which is why l^2 is called "l squared." Figure 1.4 shows this for a side of length 3 units, where the area is clearly $3 \times 3 = 9$ square units.

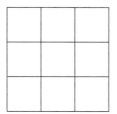

Figure 1.4: Area of a square.

Thus if a and b are the perpendicular sides of the triangle, and c is the third side, the Pythagorean theorem can be written as the equation

$$a^2 + b^2 = c^2.$$

Conversely, any triple (a, b, c) of positive numbers satisfying this equation is the triple of sides of a right-angled triangle. The story of natural numbers in geometry begins with the discovery that the equation has many solutions with natural number values of a, b, and c, and hence there are many right-angled triangles with natural number sides. The simplest has $a = 3$, $b = 4$, $c = 5$, which corresponds to the equation

$$3^2 + 4^2 = 9 + 16 = 25 = 5^2.$$

The next simplest solutions for (a, b, c) are $(5,12,13)$, $(8,15,17)$, and $(7,24,25)$, among infinitely many others, called *Pythagorean triples*. As long ago as 1800 BCE, the Babylonians discovered Pythagorean triples with values of a and b in the thousands.

The Babylonian triples appear on a famous clay tablet known as Plimpton 322 (from its museum catalog number). Actually only the b and c values appear, but the a values can be inferred from the fact that in each case $c^2 - b^2$ is the square of a natural number—something that could hardly be an accident! Also, the pairs (b, c) are listed in an order corresponding to the values of b/a, which steadily decrease, as you can see from Figure 1.5.

a	b	c	b/a
120	119	169	0.9917
3456	3367	4825	0.9742
4800	4601	6649	0.9585
13500	12709	18541	0.9414
72	65	97	0.9028
360	319	481	0.8861
2700	2291	3541	0.8485
960	799	1249	0.8323
600	481	769	0.8017
6480	4961	8161	0.7656
60	45	75	0.7500
2400	1679	2929	0.6996
240	161	289	0.6708
2700	1771	3229	0.6559
90	56	106	0.6222

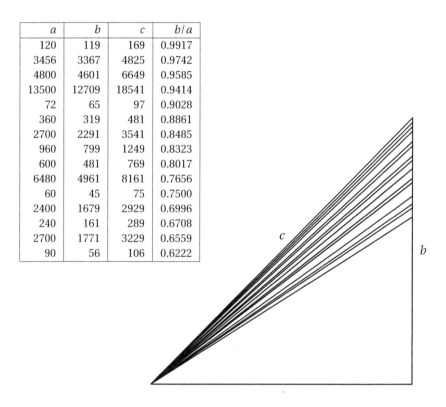

Figure 1.5: Triangles derived from Plimpton 322.

As you can also see, the slopes form a rough "scale," rather densely filling a range of angles between 30° and 45°. It looks as though the

Babylonians believed in a world of natural number ratios, rather like the Pythagoreans, and this could be an exercise like subdividing the octave by natural number ratios. But if so, there is a glaring hole in the rational geometric world: right at the top of the scale there is no triangle with $a = b$.

It is to the credit of the Pythagoreans that they alone—of all the discoverers of the Pythagorean theorem—were bothered by this hole in the rational world. They were sufficiently bothered that they tried to understand it, and in doing so discovered an *irrational* world.

Exercises

Each Pythagorean triple (a, b, c) in Plimpton 322 can be "explained" in terms of a simpler number x, given in the following table. (The numbers x are *not* in Plimpton 322 but, as we explain below, they provide a very plausible explanation of it.)

a	b	c	x
120	119	169	12/5
3456	3367	4825	64/27
4800	4601	6649	75/32
13500	12709	18541	125/54
72	65	97	9/4
360	319	481	20/9
2700	2291	3541	54/25
960	799	1249	32/15
600	481	769	25/12
6480	4961	8161	81/40
60	45	75	2
2400	1679	2929	48/25
240	161	289	15/8
2700	1771	3229	50/27
90	56	106	9/5

For each line in the table,

$$\frac{b}{a} = \frac{1}{2}\left(x - \frac{1}{x}\right).$$

1.2.1 Check that $\frac{1}{2}\left(x-\frac{1}{x}\right) = \frac{119}{120}$ when $x = 12/5$.

1.2.2 Also check that

$$\frac{b}{a} = \frac{1}{2}\left(x-\frac{1}{x}\right).$$

for three other lines in the table.

The numbers x are not only "shorter" than the numbers b/a, they are "simple" in the sense that they are built from the numbers 2, 3, and 5. For example

$$\frac{12}{5} = \frac{2^2 \times 3}{5} \quad \text{and} \quad \frac{125}{54} = \frac{5^3}{2 \times 3^3}.$$

Numbers divisible by 2, 3, or 5 were "round" numbers in the view of the Babylonians, whose number system was based on the number 60. Remnants of this number system are still in use today; for example, we divide the circle into 360 degrees, the degree into 60 minutes, and the minute into 60 seconds.

1.2.3 Check that every other fraction x in the table can be written with both numerator and denominator as a product of powers of 2, 3, or 5.

The formula $\frac{1}{2}\left(x-\frac{1}{x}\right) = \frac{b}{a}$ gives us whole numbers a and b from a rational number x. But why should there be a whole number c such that $a^2 + b^2 = c^2$? Let us see:

1.2.4 Verify by algebra that

$$\left[\frac{1}{2}\left(x-\frac{1}{x}\right)\right]^2 + 1 = \left[\frac{1}{2}\left(x+\frac{1}{x}\right)\right]^2.$$

1.2.5 Deduce from 1.2.4 that $a^2 + b^2 = c^2$, where

$$\frac{1}{2}\left(x+\frac{1}{x}\right) = \frac{c}{a}.$$

1.2.6 Check that the formula in 1.2.5 gives $c = 169$ when $x = 12/5$ (the first line of the table), and also check three other lines in the table.

1.3   Irrational Triangles

Surely the simplest triangle in the world is the one that is half a square, that is, the triangle with two perpendicular sides of equal length (Figure 1.6). If we take the perpendicular sides to be of length 1, then the hypotenuse c satisfies $c^2 = 1^2 + 1^2 = 2$, by the Pythagorean theorem. Hence c is what we call $\sqrt{2}$, the *square root* of 2.

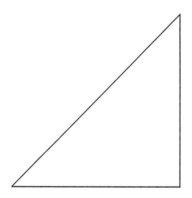

Figure 1.6: The simplest triangle.

Is $\sqrt{2}$ a ratio of natural numbers? No one has ever found such a ratio, but perhaps this is simply because we have not looked far enough. The Pythagoreans found that *no such ratio exists*, probably using some simple properties of even and odd numbers. They knew that the square of an odd number is odd, for example, and hence that an even square is necessarily the square of an even number. However, this is the easy part. The hard part is just to imagine proving that $\sqrt{2}$ is not among the ratios of natural numbers, when such ratios are the only numbers we know.

This calls for a daring method of proof known as *proof by contradiction* or *reductio ad absurdam* ("reduction to an absurdity"). To show that $\sqrt{2}$ is not a ratio of natural numbers we *suppose it is* (for the sake of argument), and deduce a contradiction. The assumption is therefore false, as we wanted to show.

In this case we begin by supposing that

$$\sqrt{2} = m/n \quad \text{for some natural numbers } m \text{ and } n.$$

We also suppose that any common factors have been cancelled from m and n. In particular, m and n are not both even, otherwise we could cancel a common factor of 2 from them both. It follows, by squaring both sides, that

$$2 = m^2/n^2$$

hence $2n^2 = m^2,$ multiplying both sides by n^2,

hence m^2 is even,   being a multiple of 2,

hence m is even,   since its square is even,

hence $m = 2l$ for some natural number l,

hence $m^2 = 4l^2 = 2n^2$ because $m^2 = 2n^2$,

hence $n^2 = 2l^2,$ dividing both sides by 2,

hence n^2 is even,

hence n is even.

But this contradicts our assumption that m and n are not both even, so $\sqrt{2}$ is *not a ratio m/n* of natural numbers. For this reason, we call $\sqrt{2}$ *irrational.*

The "Irrational" and the "Absurd"

In ordinary speech "irrational" means illogical or unreasonable—rather a prejudicial term to apply to numbers, one would think, so how can mathematicians do it without qualms? The way this came about is an interesting story, which shows in passing how accidental the evolution of mathematical terminology can be.

In ancient Greece the word *logos* covered a cluster of concepts involving speech: language, reason, explanation, and number. It is the root of our word *logic* and all the words ending in *-ology*. As we know, the Pythagoreans regarded number as the ultimate medium for explanation, so *logos* also meant ratio or calculation. Conversely, the opposite word *alogos* meant the opposite of rational, both in the general sense and in geometry, where Euclid used it to denote quantities not expressible as ratios of natural numbers.

Logos and *alogos* were translated into Latin as *rationalis* and *irrationalis*, and first used in mathematics by Cassiodorus, secretary of the

Ostrogoth king Theodoric, around 500 CE. The English words *rational* and *irrational* came from the Latin, with both the mathematical and general meaning intact.

Meanwhile, *logos* and *alogos*, because they can mean "expressible" and "inexpressible," were translated into Arabic with the slightly altered meanings "audible" and "inaudible" in the writings of the mathematician al-Khwārizmī around 800 CE. Later Arabic translators bent "inaudible" further to "dumb," from which it re-entered Latin as *surdus*, meaning "silent." Finally, *surdus* became the English word *surd* in Robert Recorde's *The Pathwaie to Knowledge* of 1551. The derived word *absurd* comes from the Latin *absurdus* meaning unmelodious or discordant, so the word has actually not strayed far from its Pythagorean origins.

However, we have come a long way from Pythagorean philosophy. It is no longer "irrational" to look for explanations outside the world of natural numbers, so there is now a conflict between the everyday use of the word "irrational" and its use in mathematics. We surely cannot stop calling unreasonable actions "irrational," so it would be better to stop calling numbers "irrational." Unfortunately, this seems to be a lost cause.

As long ago as 1585, the Dutch mathematician Simon Stevin railed against using the words "irrational" and "absurd" for numbers, but his advice has not been followed. In this broadside, Stevin avoids using absolute terms for numbers, like "irrational," by using the relative term *incommensurable* ("no common measure") for any pair of numbers not in natural number ratio. He also calls rational numbers "arithmetical." Here is a paraphrase by D. J. Struik of Stevin's words from *l'Arithmetique*, in [49, vol. IIB, p. 533].

> *That there are no absurd, irrational, irregular, inexplicable or surd numbers*
>
> It is true that $\sqrt{8}$ is incommensurable with an arithmetical number, but this does not mean it is absurd etc. ... if $\sqrt{8}$ and an arithmetical number are incommensurable, then it is as much (or as little) the fault of $\sqrt{8}$ as of the arithmetical number.

Exercises

There is one step in the irrationality proof above that perhaps needs further justification: if m^2 is even then m is even.

1.3.1 Suppose on the contrary that m is *odd*; that is, $m = 2p + 1$ for some whole number p. If so, show that

$$m^2 = 2(2p^2 + 2p) + 1,$$

which contradicts the fact that m^2 is even.

It is useful to have this nitpicking explanation why m is even when m^2 is even, because we can use the same idea to deal with the similar problem that comes up in proving $\sqrt{3}$ irrational: proving that m is a multiple of 3 when m^2 is a multiple of 3.

The *multiples of 3* are numbers $3n$, where n is a whole number. Numbers that are *not* multiples of 3 are numbers of the form $3n + 1$ and $3n + 2$ (which leave remainder 1 and 2, respectively, when divided by 3).

1.3.2 Show that
$$(3n + 1)^2 = 3(3n^2 + 2n) + 1,$$

and explain why this implies that *the square of a number that leaves remainder 1 (when divided by 3) also leaves remainder 1.*

1.3.3 It is *not* true that the square leaves remainder 2 when the number itself leaves remainder 2 (when divided by 3). Give an example.

1.3.4 Using algebra similar to that in 1.3.2, show that *the square of a number that leaves remainder 2 (when divided by 3) leaves remainder 1.*

1.3.5 Deduce from 1.3.2 and 1.3.4 that if m^2 is a multiple of 3 then so is m.

1.3.6 Use 1.3.5 to give a proof that $\sqrt{3}$ is irrational.

1.4   The Pythagorean Nightmare

The discovery of irrationality in geometry was a terrible blow to the dream of a world governed by natural numbers. The diagonal of the unit square is surely real—as real as the square itself—yet its length in units is not a ratio of natural numbers, so from the Pythagorean viewpoint it cannot be expressed by number at all. Later Greek mathematicians coped with this nightmare by developing geometry as a nonnumerical subject: the study of quantities called *magnitudes*.

Magnitudes include quantities such as length, area, and volume. They also include numbers, but, in the ancient Greek view, length does not enjoy all the properties of numbers. For example, the product of two numbers is itself a number, but the product of two lengths is *not* a length—it is a rectangle.

It is true that a rectangle with sides 2 and 3 consists of $2 \times 3 = 6$ unit squares, reflecting the fact that the product of the *numbers* 2 and 3 is the number 6. Today we exploit this parallel between area and multiplication by calling the rectangle a "2×3 rectangle." But the rectangle with sides $\sqrt{2}$ and $\sqrt{3}$ does not consist of any "number" of unit squares in the Pythagorean sense of number (Figure 1.7).

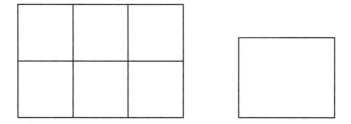

Figure 1.7: The 2×3 rectangle versus the $\sqrt{2} \times \sqrt{3}$ rectangle.

This blocks the general idea of *algebra*, where addition and multiplication are unrestricted, and it also causes mischief with the normally straightforward concept of *equality*. Two figures are shown to have equal area by cutting one figure into pieces and reassembling them to form the other. It turns out that any polygon can be "measured" in this way by a unique square, so, with some difficulty, the Greek theory of area gives the same results as ours. In fact, "equality by cutting and pasting" even gives neat proofs of a few algebraic identities. Figure 1.8 shows why $a^2 - b^2 = (a-b)(a+b)$.

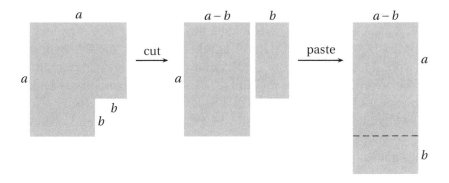

Figure 1.8: Difference of two squares: why $a^2 - b^2 = (a-b)(a+b)$.

But there is worse trouble with the concept of volume. The Greeks viewed the product of three lengths a, b, and c as a *box* with perpendicular sides a, b, and c, and they measured volumes by boxes. There are at least two problems with this train of thought.

- Volume cannot be determined by a finite number of cuts. The appropriate measure of volume is the cube, but not every polyhedron can be cut into a finite number of pieces that reassemble to form a cube. In fact, to measure the volume of a tetrahedron, it is necessary to cut it into infinitely many pieces. (See Section 4.3 for a way to do this.) The Greeks did not know it, but irrationality is the problem here too. In 1900, the German mathematician Max Dehn showed that the reason for the difficulty with the tetrahedron is that the angle between its faces is not a rational multiple of a right angle.

- We are at a loss to decide what the product of four lengths means, because we cannot visualize space with more than three dimensions.

This led to a split between geometry and number theory in Greek mathematics, ultimately to the detriment of geometry. The split is clear in Euclid's *Elements*, the most influential mathematics book of all time.

Written around 300 BCE, the *Elements* remained the basis of mathematics education in Europe until the late nineteenth century, and it still sells well today. (I recently saw a copy in an airport bookstore in Los Angeles.)

The *Elements* is divided into 13 Books, filling three large volumes in the standard English edition. The first six books cover magnitude in general, and they contain the basic theorems about length, area, and angle. For example, the Pythagorean theorem is Proposition 47 of Book I (and also Proposition 31 of Book VI, where a more sophisticated proof is given).

The theory of natural numbers as we know it—divisibility, common divisors, and prime numbers—begins only in Book VII, a point that was seldom reached in elementary mathematics instruction. Thus the panic over the irrational held up the development of algebra and number theory. The latter disciplines did not really take off until the sixteenth century, when Europeans finally overcame their fear of multiplying more than three quantities together. We take up this story in the next chapter.

In fact, not many readers of the *Elements* made it through Book V, a very deep and subtle book that explains the relationship of magnitudes to natural numbers. Book V contains the key to treating magnitudes *as* numbers, and hence enabling calculation with the irrational. However, it falls short of actually doing this, for reasons that will be explained in the next section.

Exercises

Draw a square with sides $a + b$ and divide it into four parts: a square of side a, a square of side b, and two rectangles.

1.4.1 Explain why the four parts of the big square have the areas a^2, ab, ab, b^2.

1.4.2 Deduce, by summing these areas, that $(a + b)^2 = a^2 + 2ab + b^2$.

1.4.3 Find a similar picture that shows $(a - b)^2 = a^2 - 2ab + b^2$, and explain why.

1.5   Explaining the Irrational

To the mind's eye, there is no apparent difference between lines with no common measure, such as the side and diagonal of a square. Each can be constructed from the other, and equally easily. This is the point of Simon Stevin's defense of irrational numbers in his *l'Arithmetique* of 1585. As we saw in Section 1.3, he objected to the very idea of calling numbers "irrational," because all numbers are equally concrete from a geometric point of view. But today we still use the term "rational," rather than Stevin's "arithmetical," so we are stuck with the term "irrational numbers" and the problem is to explain what they are. How can we perceive the difference between rational and irrational numbers? And how can an irrational number, such as $\sqrt{2}$, be grasped at all?

The key is to compare irrationals with what we know, the rational numbers. This simple but profound idea began with Eudoxus, around 350 BCE, and it is developed at length in Book V of Euclid's *Elements*. An irrational line, such as the diagonal of the unit square, is not any rational number of units in length, but we know (for example) that it is greater than one unit and less than two. In fact, we can say precisely which rational multiples of the unit are less than the diagonal and which are greater. Eudoxus realized that this is all we need to know: *an irrational is known by the rational numbers less than it, and the rational numbers greater than it.*

We can make this idea even more down-to-earth by using only the most familiar type of rational numbers, the *decimal fractions*. These are the numbers we write with a decimal point and a finite number of digits, such as 1.42, which stands for the rational number $\frac{142}{100}$. Not all rational numbers are decimal fractions, but there are enough decimal fractions to pin down any irrational number: *an irrational is known by the decimal fractions less than it, and the decimal fractions greater than it.*

Here is how this mathematical version of "being known by the company you keep" applies to the irrational $\sqrt{2}$. We know something about $\sqrt{2}$ if we know that

$$1 < \sqrt{2} < 2;$$

we know more if we know that

$$1.4 < \sqrt{2} < 1.5;$$

we know even more if we know that

$$1.41 < \sqrt{2} < 1.42;$$

still more if we know that

$$1.414 < \sqrt{2} < 1.415,$$

and so on.

We know $\sqrt{2}$ *completely* if we know, for any decimal fraction d, whether $d < \sqrt{2}$ or $d > \sqrt{2}$. This is because no other quantity lies at the same position, above and below the same decimal fractions, as $\sqrt{2}$. Any other quantity q must differ from $\sqrt{2}$ by a certain amount, greater than $0.00\cdots01$ say. But in that case q will lie above some decimal fraction $d + 0.00\cdots01$ that $\sqrt{2}$ lies below, or vice versa. Thus *the gap in the rationals occupied by $\sqrt{2}$ is "point-sized"*—no other quantity falls into it.

To locate the gap we compare $\sqrt{2}$ with its neighbors among the decimal fractions, looking in turn at the decimal fractions with $1, 2, 3, \ldots$ digits, as we started to do above. But now we let the number of digits increase indefinitely, so that the width of the gap shrinks towards zero.

The neighbors below $\sqrt{2}$ (since their squares are less than 2) are

> 1
>
> 1.4
>
> 1.41
>
> 1.414
>
> 1.4142
>
> 1.41421
>
> 1.414213
>
> 1.4142135
>
> 1.41421356
>
> 1.414213562
>
> 1.4142135623
>
> 1.41421356237
>
> 1.414213562373
>
> $\vdots$

and those above it (since their squares are greater than 2) are

$$\vdots$$

1.414213562374

1.41421356238

1.4142135624

1.414213563

1.41421357

1.4142136

1.414214

1.41422

1.4143

1.415

1.42

1.5

2,

so the position of $\sqrt{2}$ between these two sets of decimal fractions is described exactly by an *infinite decimal* that begins

$$1.414213562373\cdots.$$

Today we often take this infinite decimal to *be* $\sqrt{2}$, and indeed infinite decimals are probably the most concrete way to represent irrationals of all kinds. Stevin briefly introduced infinite decimals in his book *De Thiende* (The Tenth) of 1585. It is possible to add and multiply infinite decimals, so infinite decimals not only look like numbers, but also behave like them.

The Greeks did not use infinite decimals to represent irrationals, because they did not even have finite decimals. It is unlikely they would have accepted infinite expressions as numbers in any case, since they did not believe in infinite objects. They were willing to accept infinite *processes*, that is, processes without end, such as listing the natural numbers $1, 2, 3, \ldots$. But they did not accept infinite objects, since an infinite object would seem to be the impossible result of ending a process that does not end. Yet infinite processes arise in any attempt to understand irrationals, and in the case of $\sqrt{2}$ the Greeks found a process that actually gives a better understanding than the infinite decimal. We study this infinite process in Section 1.6.

If You Can Read This, Thank an English Teacher

There is a reasonably simple algorithm for computing infinite decimals of square roots, which was actually taught in schools in the early part of the twentieth century. In my schooldays it was no longer on the syllabus, but I happened to meet it one day in seventh or eighth grade when the mathematics teacher was away sick. The English teacher, Mrs. Burke, substituted for him, and taught us something she remembered from her own schooldays: the square root algorithm. As the successive digits of $\sqrt{2}$ emerged, I realized with amazement and delight that there are *mysteries* in mathematics.

No matter how many digits we examine, say the first 40,

$$1.4142135623730950488016887242209698078569\cdots,$$

there is no telling what comes next. Mrs. Burke told us that there is no known pattern to the digits, and her statement is probably still accurate. The sequence of digits seems somehow random, but we can't prove *that* either. We don't know whether each digit occurs with the same frequency; we don't even know if a particular digit, say 7, occurs infinitely often. Thus we don't really "know" the infinite decimal for $\sqrt{2}$, only a process for computing any finite part of it.

This is how I first got interested in mathematics, and irrational numbers in particular. Getting to "know" the infinite decimal for $\sqrt{2}$ may be too discouraging for some—mathematicians generally prefer a problem that gives them some chance of success—but it is a reminder of how much remains to be understood. And since the infinite decimal is so completely baffling, we are encouraged to look at different approaches to $\sqrt{2}$, in the hope of finding enlightenment elsewhere. Surely $\sqrt{2}$ is a simple irrational number, so there should be a simple way of looking at it.

Exercises

Let us call 1.4 the *neighbor below* $\sqrt{2}$ with one decimal digit because
it is the *largest* such fraction below $\sqrt{2}$. Similarly, we will call 1.5 the
neighbor above $\sqrt{2}$ with one decimal digit because it is the *smallest*
such fraction above $\sqrt{2}$.

The neighbors with $2, 3, \ldots$ decimal digits are defined similarly. 1.41
is the neighbor below $\sqrt{2}$ with two decimal digits because 1.41 is below
$\sqrt{2}$ but 1.42 is not. It follows that 1.42 is the neighbor above $\sqrt{2}$ with
two decimal digits. 1.414 is the neighbor below $\sqrt{2}$ with three decimal
digits because 1.414 is below $\sqrt{2}$ but 1.415 is not, and so on.

The following exercises ask you to find neighbors of some other irrational numbers. You may use a calculator if you wish.

1.5.1 Find the decimal fractions with $1, 2, 3, 4$ digits that are neighbors
below and above $\sqrt{2}/2$.

1.5.2 Find the decimal fractions with $1, 2, 3, 4$ digits that are neighbors
below and above $3\sqrt{2}$.

The numbers $\sqrt{2}/2$ and $3\sqrt{2}$ are also irrational, as we see from the
following more general fact.

1.5.3 If x is any irrational number, then $x/2$ and $3x$ are also irrational.
Explain why. (Hint: If the number $3x$ is rational, what can you
conclude about x?)

1.6   The Continued Fraction for $\sqrt{2}$

A more enlightening process than the infinite decimal algorithm would
relate $\sqrt{2}$ to a *predictable* infinite sequence of numbers, and such a
process actually exists: the famous *Euclidean algorithm*. The algorithm is named after Euclid because its first known appearance is in
the *Elements*, Book VII, but there is evidence for its earlier use in the
investigation of irrationals.

The object of the Euclidean algorithm is to find a *common measure*
of two magnitudes a and b. We say that c is a common measure of a
and b if

$$a = mc \quad \text{and} \quad b = nc \quad \text{for some natural numbers } m \text{ and } n.$$

The key to the algorithm is the fact that

$$a - b = (m - n)c,$$

so *any common measure c of a and b is also a measure of their difference*. If *a* is greater than *b*, we can therefore replace the problem of finding a common measure of *a* and *b* by the "smaller" problem of finding a common measure of $a - b$ and *b*. This is precisely what the algorithm does. The algorithm seeks a common measure by "continually subtracting the lesser magnitude from the greater" (as Euclid put it).

When a common measure exists the Euclidean algorithm will find it in a finite number of steps. In particular, when *a* and *b* are positive natural numbers the algorithm gives what is called the *greatest common divisor* of *a* and *b*. This concept is very important in the theory of natural numbers, and we pursue it further in Chapter 7.

Right now, we are interested in the opposite situation, where *a* and *b* have no common measure. In this case the algorithm necessarily runs forever, but perhaps in a way that enables us to understand *why* it runs forever. We are in luck when $a = \sqrt{2}$ and $b = 1$. Here, the Euclidean algorithm runs forever because it becomes *periodic*: it returns to a situation it has been in before. Unlike the infinite decimal process for $\sqrt{2}$, the Euclidean algorithm on $\sqrt{2}$ and 1 has infinite behavior of the simplest possible kind.

To show periodic behavior as clearly as possible, I apply the Euclidean algorithm instead to the numbers $a = \sqrt{2} + 1$ and $b = 1$, where periodicity kicks in a little earlier. Also, I present the algorithm in a geometric form. Two magnitudes *a* and *b* will be represented by the greater and lesser sides of a rectangle, and the lesser *b* is subtracted from the greater *a* by cutting off a square of side *b* (Figure 1.9).

When $a = \sqrt{2} + 1$ and $b = 1$ we can subtract *b* from *a* twice before getting a magnitude less than *b*. The result is a rectangle with greater side 1 and lesser side $\sqrt{2} - 1$, as shown in Figure 1.10. I claim *this is a repetition of the original situation* because *the new rectangle is the same shape as the old.*

To see why, we calculate the ratio $\dfrac{\text{long side}}{\text{short side}}$, which represents the

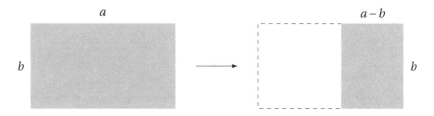

Figure 1.9: Subtracting the lesser from the greater of two magnitudes.

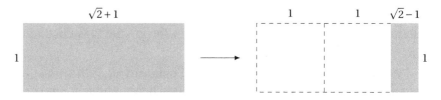

Figure 1.10: Subtracting 1 twice from $\sqrt{2}+1$.

shape of the rectangle. For the old rectangle,

$$\frac{\text{long side}}{\text{short side}} = \frac{\sqrt{2}+1}{1} = \sqrt{2}+1.$$

For the new rectangle,

$$\frac{\text{long side}}{\text{short side}} = \frac{1}{\sqrt{2}-1}$$

$$= \frac{1}{\sqrt{2}-1} \frac{\sqrt{2}+1}{\sqrt{2}+1} \quad \text{multiplying top and bottom by } \sqrt{2}+1$$

$$= \frac{\sqrt{2}+1}{(\sqrt{2})^2 - 1^2} \quad \text{because } (\sqrt{2}-1)(\sqrt{2}+1) = (\sqrt{2})^2 - 1^2,$$

by the "difference of two squares" from Section 1.4

$$= \frac{\sqrt{2}+1}{2-1} = \sqrt{2}+1.$$

Thus the new rectangle has the same shape as the old, so as we continue to run the Euclidean algorithm, the same thing will happen over and over: two squares will be cut off the new rectangle, producing yet another rectangle of the same shape, and so on. The algorithm will not terminate because it keeps reproducing the same geometric situation.

It is also clear that the Euclidean algorithm will not terminate if it starts with the pair $\sqrt{2}$ and 1. The only change is that just one square is cut off the initial rectangle. After that, rectangles with sides in the ratio $\sqrt{2}+1:1$ recur forever.

Since the Greeks knew that $\sqrt{2}$ is irrational, they knew that the Euclidean algorithm does not terminate when applied to $\sqrt{2}$ and 1. Indeed, the *Elements*, Book X, Proposition 2 actually states nontermination of the algorithm as a criterion for irrationality. What is not so clear is whether the Greeks observed the *periodicity* of the algorithm on $\sqrt{2}$ and 1. Many people believe they did, but the *Elements* does not mention it, and the other likely sources of this information are lost.

One thing is clear. It is a lot easier to display the periodicity of $\sqrt{2}$ using notation for fractions, which the Greeks did not have. Again, it is better to use $\sqrt{2}+1$ for purposes of demonstration, because the periodicity starts earlier. We start by subtracting 1 twice from $\sqrt{2}+1$, as the algorithm requires. This produces the remainder $\sqrt{2}-1$, which equals $\frac{1}{\sqrt{2}+1}$ because $(\sqrt{2}-1)(\sqrt{2}+1)=1$, as found above. Thus

$$\sqrt{2}+1 = 2+(\sqrt{2}-1) = 2+\frac{1}{\sqrt{2}+1}.$$

This says that the $\sqrt{2}+1$ in the bottom of the fraction on the right side equals the *whole* right side. We can therefore replace $\sqrt{2}+1$ by $2+\frac{1}{\sqrt{2}+1}$, and keep on doing this to our hearts' content:

$$\sqrt{2}+1 = 2+\frac{1}{\sqrt{2}+1} = 2+\frac{1}{2+\frac{1}{\sqrt{2}+1}} = 2+\frac{1}{2+\frac{1}{2+\frac{1}{\sqrt{2}+1}}} \quad \cdots$$

The logical conclusion of this process is an infinite periodic fraction for $\sqrt{2}+1$, called its *continued fraction*:

$$\sqrt{2}+1 = 2+\cfrac{1}{2+\cfrac{1}{2+\cfrac{1}{2+\cfrac{1}{2+\cfrac{1}{2+\ddots}}}}}$$

Finally, we get the continued fraction for $\sqrt{2}$ by subtracting 1 from both sides:

$$\sqrt{2} = 1 + \cfrac{1}{2 + \cfrac{1}{2 + \cfrac{1}{2 + \cfrac{1}{2 + \cfrac{1}{2 + \cdots}}}}}$$

The infinite recurrence of 2s produced by the Euclidean algorithm is reflected by the infinite sequence of 2s in the continued fraction.

Periodicity in Continued Fractions and Decimals

You may be wondering whether it is valid to expel $\sqrt{2}$ from the right side of the equations as we did above, by letting the bottom line of the fraction descend to infinity. But if we believe that the continued fraction is meaningful we can verify that its value is $\sqrt{2} + 1$ as follows. Let

$$x = 2 + \cfrac{1}{2 + \cfrac{1}{2 + \cfrac{1}{2 + \cfrac{1}{2 + \cfrac{1}{2 + \cdots}}}}}$$

Then x is a positive number (greater than 2 in fact), and since the term under the 1 on the right side equals the whole right side, we also have

$$x = 2 + \frac{1}{x}.$$

Multiplying both sides by x, and rearranging, we get the equation

$$x^2 - 2x - 1 = 0.$$

This quadratic equation has two solutions, $1 + \sqrt{2}$ and $1 - \sqrt{2}$, but only the first one is positive, hence it is the value of the continued fraction.

A similar argument shows that any periodic continued fraction is the solution of a quadratic equation, which means periodicity is quite rare. In infinite decimals, periodicity is even more special: it occurs only for rational numbers. The reason for this can be seen by taking a random example, such as

$$x = 0.235717171717171\cdots$$

First we shift the decimal point three places (the length of the segment 235 that comes before the periodic part) by multiplying by 1,000:

$$1000x = 235.717171717171\cdots$$

Then we shift by another two places (the length of the "period" 71) by multiplying by 100:

$$100000x = 23571.717171717171\cdots$$

Finally we subtract $1000x$ from $100000x$, to cancel the parts after the decimal point:

$$100000x - 1000x = 23571 - 235, \quad \text{that is,} \quad 99000x = 23336,$$

so x is the rational number $\frac{23336}{99000}$.

The converse result, that the decimal of any fraction is periodic, is well-known in certain instances. For example,

$$\frac{1}{3} = 0.33333333\ldots \quad \text{and} \quad \frac{1}{6} = 0.16666666\ldots,$$

but a general explanation depends on the division process learned in school, namely *division with remainder*, a process we will revisit in Chapter 7. Successive decimal digits come from remainders of successive divisions by the denominator of the fraction, and the remainders eventually repeat because they are all less than the denominator.

Exercises

1.6.1 Use the division process to show that

$$\frac{1}{7} = 0.142857142857142857\ldots.$$

Confirm this result by subtracting $x = 0.142857142857142857\ldots$ from $1000000x$.

Another geometrically interesting irrational number is the *golden ratio* $\frac{1+\sqrt{5}}{2}$, which is the ratio $\frac{\text{long side}}{\text{short side}}$ of the *golden rectangle*, which has width $\frac{1+\sqrt{5}}{2}$ and height 1.

1.6.2 Use a calculator to check that the golden ratio is approximately 1.618.

1.6.3 Sketch a golden rectangle and show that the result of removing a square of side 1 is a rectangle of width $\frac{\sqrt{5}-1}{2}$ and height 1.

1.6.4 Show that

$$\frac{1}{(\sqrt{5}-1)/2} = \frac{1+\sqrt{5}}{2},$$

that is, *the new rectangle is the same shape as the old.*

1.6.5 Deduce from 1.6.4 that the golden ratio $\frac{1+\sqrt{5}}{2}$ equals the continued fraction

$$1 + \cfrac{1}{1 + \cfrac{1}{1 + \cfrac{1}{1 + \cfrac{1}{1 + \cfrac{1}{1 + \cdots}}}}}$$

1.6.6 Confirm the result of 1.6.5 by letting

$$x = 1 + \cfrac{1}{1 + \cfrac{1}{1 + \cfrac{1}{1 + \cfrac{1}{1 + \cfrac{1}{1 + \cdots}}}}}$$

Explain why $x = 1 + \frac{1}{x}$ and show that the positive solution of this equation is $x = \frac{1+\sqrt{5}}{2}$.

1.7 Equal Temperament

Now it is time to return to the music question raised in Section 1.1: can a sum of fifths equal a sum of octaves? The answer is no, because summing m fifths corresponds to multiplying frequency by $(3/2)^m$, while summing n octaves corresponds to multiplying frequency by 2^n. If the two sums of intervals are equal, then

$$\left(\frac{3}{2}\right)^m = 2^n,$$

and therefore, multiplying both sides by 2^m,

$$3^m = 2^m \times 2^n.$$

But this is impossible, because the left side is odd (being a product of odd numbers) and the right side is even (being a product of even numbers). Thus the octave and the fifth have no "common measure" and the Pythagorean attempt to divide the octave into natural steps using rational numbers was doomed from the start—though the Pythagoreans may never have noticed.

If one wants a scale with the sweetest harmonies, given by the octave and the fifth, then it is impossible to include both by dividing the octave into equal steps. An *approximation* to such a scale is possible, since the sum of 12 fifths is close to seven octaves. But (as mentioned in Section 1.1) the difference between 12 fifths and seven octaves is a noticeable fraction of a step, so some compromise is necessary. If the perfect fifth is a note in the scale, then some steps in the scale that are supposed to be equal will not be. If we insist on equal steps, then the fifth note of the scale will be imperfect—its frequency ratio with the first note will not be exactly 3:2.

This problem bedevilled music in both the East and West. In China, music was generally based on a different scale (dividing the octave into five steps instead of seven—corresponding to the black keys on the piano), but the Chinese also tried to build their scale from a cycle of 12 perfect fifths, and ran into the same problem. Amazingly, the compromise solution that builds the scale from 12 equal parts of an octave ("semitones") was proposed in the East and West almost simultaneously: by Zhu Zaiyu in China and by Simon Stevin in the Netherlands. Their discoveries are generally given the dates 1584 and 1585, respectively, but neither is known exactly, so it seems fair to consider the discoveries independent. (An additional fact, which makes coincidence almost unbelievable, is that the 1580s were also the decade when Jesuit missionaries arrived in China, bringing western mathematics to the region for the first time.)

Both Zhu and Stevin realized that adding 12 equal semitones to form an octave amounts to doing 12 multiplications, by the same number, to achieve multiplication by 2. Therefore, one semitone corresponds to multiplication by the 12th root of 2, or $2^{1/12}$ as Stevin wrote it. In both eastern and western scales the crucial note is seven semi-tones above the first, hence at a frequency ratio of

$$\left(2^{1/12}\right)^7 = 2^{7/12} = 1.49831\cdots$$

This is certainly close to the perfect fifth ratio of 1.5, though some people can hear the difference. The advantage of this imperfect fifth, however, is that 12 of them make exactly seven octaves, and a cycle of 12 fifths hits all notes of the new scale exactly.

The system of equal semitones, or *equal temperament*, obviously has great simplicity and mathematical beauty. However, it was not immediately popular with musicians. In western music it did not become widespread until the nineteenth century. Bach's *Well-Tempered Clavier* of 1722 is often thought to be an advertisement for equal temperament, but more likely it was intended to show off an older system. There are many systems of "just temperament" that retain perfect fifths while coming closer to equal semitones than the Pythagorean scale.

For a modern mathematician, of course, equal temperament is interesting because the basic ratio $2^{1/12}$ is *irrational*. It must be, because its sixth power is the irrational number $\sqrt{2}$, while powers of rational numbers are rational. Thus *equal temperament rejects all integer ratios in music except the ratio 2:1 for the octave.* For a long time, the Chinese do not seem to have been aware of irrational numbers. According to Joseph Needham's *Science and Civilisation in China* [40, vol. III, p. 90]:

> The Chinese mathematicians ... seem to have been nei-
> ther attracted nor perplexed by irrationals, if indeed they
> appreciated their separate existence.

Zhu Zaiyu may be an exception to this characterization. In 1604, he wrote a *New Explanation of the Theory of Calculation* in which he derived values of the roots of 2 needed for equal temperament. He was so attracted to $\sqrt{2}$ that he used nine abacuses to compute it to 25-digit accuracy!

As we know from Section 1.3, Stevin appreciated the existence of irrationals, and aggressively asserted their equal rights with rational numbers. In his *Theory of the Art of Singing* he seized the fresh opportunity to mock the unbelievers in irrational numbers. He even claimed that the equal-tempered fifth must *sound* sweet because $2^{7/12}$ is such a wonderful number!

Now someone might wonder, according to the ancient view, how the sweet sound of the fifth could consist in so unspeakable, irrational, and inappropriate a number. To this we might give a detailed answer. However, … it is not our intention to teach those with the inexpressible irrationality and inappropriateness of such a misunderstanding the expressibility, rationality, appropriateness, and natural wonderful perfection of these numbers … [1]

Figure 1.11: Frets on a guitar.

This is Pythagoreanism reborn! "All is number" again, but the meaning of "number" has been expanded to include irrationals.

In conclusion, we should mention that both Zhu and Stevin saw musical ratios the way Pythagoras did—as ratios of string lengths—because they were writing a few decades before Beeckman discovered that pitch corresponds to frequency of vibration. Thus the most direct realization of their idea today is seen in the placement of frets on an instrument such as a guitar (see Figure 1.11). Sliding the finger from one fret to the next changes the length of the vibrating string by a factor of $2^{1/12}$.

Exercises

It is mentioned above that $2^{1/12}$ is irrational because its sixth power is $\sqrt{2}$ (which we know to be irrational) and powers of rational numbers are irrational. Here is why:

---

[1] Translation by Adriaan Fokker in *The Principal Works of Simon Stevin*, vol. V, p. 441. (Modified by the author with the assistance of Hendrik Lenstra.)

1.7.1 Explain why $\left(2^{1/12}\right)^6 = \sqrt{2}$.

1.7.2 Suppose r is rational, that is $r = m/n$ for some whole numbers m and n. If k is a whole number, what is r^k, and why is it rational?

We can also check that intervals of $1, 2, 3, 4, 5, 6, 7, 8, 9, 10$ and 11 semitones correspond to irrational frequency ratios. One semitone corresponds to the ratio $2^{1/12} : 1$, and we have already seen that $2^{1/12}$ is irrational.

1.7.3 $\left(2^{1/12}\right)^2$ is irrational because its cube is $\sqrt{2}$. Explain.

1.7.4 $\left(2^{1/12}\right)^3$ is irrational because its ... what?

1.7.5 $\left(2^{1/12}\right)^4 = 2^{1/3}$, the cube root of 2. Give a proof that the cube root of 2 is irrational along the same lines as the proof in Section 1.3.

1.7.6 $\left(2^{1/12}\right)^5$ is irrational because its cube is $2 \times 2^{1/3}$, which is irrational because $2^{1/3}$ is. Explain.

Chapter 2

The Imaginary

Preview

In Chapter 1 we saw why the numbers $1, 2, 3, \ldots$, used for counting and basic arithmetic, do not meet all the needs of geometry. To measure line segments we need irrational numbers such as $\sqrt{2}$, and indeed a continuous sequence of numbers that fill the spaces between $1, 2, 3, \ldots$ with something like a line itself.

We get a full line, called the *real number line* $\mathbb{R}$, by extending the continuous sequence of numbers backward through $0, -1, -2, -3, \ldots$, thus balancing each positive number x with its *negative*, $-x$. This gives a line, infinite in both directions, whose points can be added, subtracted, multiplied, and divided without restriction (except division by zero).

The name "real" of course suggests that no other numbers really exist, but algebra asks for more. Algebra is supposed to solve equations, such as

$$x^3 - 15x = 4.$$

This equation has a real solution, but there is something strange about it: according to the formula for solving such equations, the solution of $x^3 - 15x = 4$ involves $\sqrt{-1}$. *Such a solution seems* impossible, *because no real number has square equal to* -1. *Indeed numbers such as* $\sqrt{-1}$ *were once called "impossible," and they are called "imaginary" even today.*

To reconcile the real solution of $x^3 - 15x = 4$ with the "impossible"

33

solution given by the formula, mathematicians had to believe that it is meaningful to calculate with "impossible" numbers. The success of "calculating with the impossible" led to $\sqrt{-1}$ being accepted as a new kind of number, now called a *complex number*.

2.1 Negative Numbers

Negative numbers represent the result of subtracting more than you have, and hence they are familiar to all people who use money. We should also not forget zero, which results from subtracting exactly what you have:

$$0 = n - n.$$

The negative numbers $-n$ can then be viewed as *mirror images* of the positive numbers n, obtained by subtracting positive numbers from zero:

$$-n = 0 - n.$$

The negatives of the natural numbers $1, 2, 3, 4, \ldots$ are called the *negative integers* $-1, -2, -3, -4, \ldots$, and together with 0 and $1, 2, 3, 4, \ldots$ they make up the *integers*. By inserting negative rational numbers between negative integers, and filling the gaps between the negative rational numbers with negative irrational numbers we get a nice symmetric *real number line* $\mathbb{R}$, extending indefinitely to right and left:

The negative numbers are familiar, as I said, to all people who use money (also to people who live in cold climates where the temperature goes below zero). However, familiarity with negative numbers extends only as far as addition and subtraction. When the temperature goes up three degrees we add 3; when the temperature goes down three degrees we subtract 3, if necessary subtracting from zero and getting a negative temperature. Likewise, when we owe three dollars we subtract 3 from our account, if necessary subtracting from zero and getting a negative balance, that is, a *debt*.

Doubts arise when a mathematician turns up and wants to *multiply* negative numbers. Many people will say it is nonsensical to multiply numbers that were intended only for addition and subtraction. For example, what is -1 times -1? However, nonsense can be avoided by demanding the *rule of law* in arithmetic: laws that apply to positive numbers should also apply to the negative.

When we extend the rule of law to negative numbers, we find there is exactly one reasonable value for -1 times -1, and in fact for every product involving negative numbers. The relevant law for evaluating products, is called the *distributive law* and it reads simply:

$$a(b + c) = ab + ac.$$

This law is certainly true for all positive numbers, though we are seldom conscious of using it. (And occasionally, people may fail to recognize that the distributive law is true, as the following experience shows. With my wife and another couple, I visited a restaurant that was offering a 25% discount. The waiter calculated our discount, *not* by multiplying our total bill by 1/4, but by multiplying the bill for each couple by 1/4, and adding the results.)

The idea of using known laws for positive numbers to govern the behavior of negative (or other) numbers became current among mathematicians in Great Britain and Germany in 1830. But, as far as I know, the first attempt to explain products of negative numbers with the help of the distributive law is in the 1585 *Arithmetic* of Simon Stevin. On page 166 of that work he expands $(8 - 5)(9 - 7)$ using the distributive law, pointing out that to get the correct answer 6 we must assume that $(-5)(-7) = 35$.

With $c = 0$ in $a(b + c)$, the distributive law explains why multiplication by zero gives zero, as follows:

$$ab = a(b + 0) \quad \text{because } b = b + 0,$$
$$= ab + a \cdot 0 \quad \text{by the distributive law,}$$
$$\text{hence} \quad 0 = a \cdot 0 \quad \text{subtracting } ab \text{ from both sides.}$$

Then, with $a = -1$, $b = 1$, $c = -1$ the law explains what $(-1)(-1)$ is:

$$0 = (-1) \cdot 0 \qquad \text{because } a \cdot 0 = 0 \text{ for any } a,$$
$$= (-1)(1 + (-1)) \qquad \text{because } 0 = 1 + (-1),$$
$$= (-1) \cdot 1 + (-1)(-1) \qquad \text{by the distributive law,}$$
$$= -1 + (-1)(-1) \qquad \text{because } a \cdot 1 = a \text{ for any } a,$$

hence $\quad 1 = (-1)(-1) \qquad$ adding 1 to both sides.

It easily follows that $(-a)^2 = a^2$ for any positive number a, so *the square of any number—positive, negative, or zero—is not negative.* In particular, -1 *is not the square of any number on the number line.* For this reason, numbers on the number line are called *real* (and the line is called $\mathbb{R}$): to contrast them with apparently *unreal, impossible,* or *imaginary* numbers whose squares are negative.

The Distributive Law in Greek Geometry

The ancient Greek interpretation of products as areas also satisfies the distributive law. If $b + c$ is the sum of lengths b and c, then rectangle $a(b + c)$ is clearly the sum of the rectangles ab and ac (Figure 2.1).

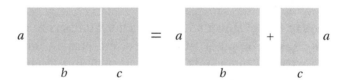

Figure 2.1: The distributive law for areas: $a(b + c) = ab + ac$.

Euclid proves special cases of the distributive law in the *Elements*, Book II, Propositions 2 and 3. He uses them to prove a geometric equivalent of the formula for the square of a sum: $(a + b)^2 = a^2 + 2ab + b^2$ (Figure 2.2).

Artmann [3, p. 63] points out that much the same figure appears on Greek coins from 404 BCE—see Figure 2.3, which is from Wikimedia. Thus the formula $(a + b)^2 = a^2 + 2ab + b^2$ was common currency, so to speak, 100 years before Euclid! It is not hard to imagine that the Greeks also considered distributive laws involving subtraction, such as $a(b - c) = ab - ac$ (Figure 2.4).

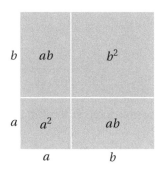

Figure 2.2: Euclid's version of $(a+b)^2 = a^2 + 2ab + b^2$.

Figure 2.3: Greek coin illustrating $(a+b)^2$

.

If so, they could have found the formula $(a-b)^2 = a^2 - 2ab + b^2$ (Figure 2.5), though they would no doubt have balked at setting $a = 0$ to get $(-b)^2 = b^2$.

(The initial figure is a square of side a plus a square of side b at its bottom left, hence its area is $a^2 + b^2$. From this we subtract the two dashed $a \times b$ rectangles, leaving the gray square of side $a - b$.)

Exercises

2.1.1 Prove $(a+b)^2 = a^2 + 2ab + b^2$ by expanding $(a+b)(a+b)$. Where did you use the distributive law?

2.1.2 Use the equation $(a+b)^2 = a^2 + 2ab + b^2$ and the distributive law again to prove

$$(a+b)^3 = a^3 + 3a^2b + 3ab^2 + b^3.$$

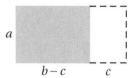

$$a$$

$$b - c \qquad c$$

Figure 2.4: Geometric version of $a(b - c) = ab - ac$.

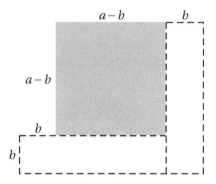

Figure 2.5: Geometric version of $(a - b)^2 = a^2 - 2ab + b^2$.

2.1.3 Using the result of 2.1.2, find $(a + b)^4$.

2.2 Imaginary Numbers

From our experience thus far, numbers with negative squares seem completely uncalled for. Why not just say, as we did in the previous section, that no real number has a negative square and leave it at that? If someone asks us to solve the equation

$$x^2 = -1,$$

we are perfectly entitled to say that there is no solution. It is similar with any quadratic equation

$$ax^2 + bx + c = 0.$$

There is a solution for some values of a, b, and c, but not for others, and it is easy to tell which case is which. No real solution exists when a square is required to be negative.

To see why, recall how you learned to solve this equation in high school. First divide both sides by a to produce the equation

$$x^2 + \frac{b}{a}x + \frac{c}{a} = 0.$$

Then notice that $x^2 + \frac{b}{a}x$ can be *completed to the square* $\left(x + \frac{b}{2a}\right)^2$ by adding $\frac{b^2}{4a^2}$, because the square of the sum $x + \frac{b}{2a}$ is

$$\left(x + \frac{b}{2a}\right)^2 = x^2 + 2\frac{b}{2a}x + \left(\frac{b}{2a}\right)^2 = x^2 + \frac{b}{a}x + \frac{b^2}{4a^2}.$$

We therefore add $\frac{b^2}{4a^2}$ to both sides of the equation $x^2 + \frac{b}{a}x + \frac{c}{a} = 0$, obtaining

$$x^2 + \frac{b}{a}x + \frac{b^2}{4a^2} + \frac{c}{a} = \frac{b^2}{4a^2},$$

$$\text{that is,} \quad \left(x + \frac{b}{2a}\right)^2 + \frac{c}{a} = \frac{b^2}{4a^2},$$

$$\text{and hence} \quad \left(x + \frac{b}{2a}\right)^2 = \frac{b^2}{4a^2} - \frac{c}{a} = \frac{b^2 - 4ac}{4a^2}.$$

Taking square roots of both sides gives

$$x + \frac{b}{2a} = \pm\frac{\sqrt{b^2 - 4ac}}{2a}.$$

Finally, subtracting $\frac{b}{2a}$ from both sides, we get the familiar *quadratic formula*:

$$x = \frac{-b \pm \sqrt{b^2 - 4ac}}{2a}.$$

This formula for the solution x involves the square root of $b^2 - 4ac$, which can be negative. But $b^2 - 4ac$ is negative precisely when the square

$$\left(x + \frac{b}{2a}\right)^2 = \frac{b^2 - 4ac}{4a^2}$$

is negative. Hence in this more general case we are also entitled to say that there is no solution.

The quantity $b^2 - 4ac$ is called the *discriminant* of the quadratic equation because it discriminates between the equations with real solutions and those without. A negative value of $b^2 - 4ac$ tells us to ignore the quadratic formula because there is no solution. Calling solutions "imaginary" when $b^2 - 4ac < 0$ is merely a highbrow way of saying "no solution" ... or is it?

By a strange quirk of history, mathematicians first found "imaginary solutions" that cannot be ignored in formulas for solving *cubic* equations. The formulas were found through heroic efforts of Italian mathematicians del Ferro and Tartaglia in the early sixteenth century. They appeared in the book *Ars Magna* (Great Art) by Cardano (1545), who fully appreciated their revolutionary nature. He wrote [6, p. 8]:

> In our own days Scipione del Ferro of Bologna has solved
> the case of the cube and first power equal to a constant,
> a very elegant and admirable accomplishment. Since this
> art surpasses all human subtlety and the perspicuity of mor-
> tal talent and is a truly celestial gift and a very clear test of
> the capacity of men's minds, whoever applies himself to
> it will believe that there is nothing that he cannot under-
> stand.

Exercises

The mathematician al-Khwārizmī, whom we mentioned in Section 1.3, solved the quadratic equation $x^2 + 10x = 39$ by literally "completing the square" in the following steps.

2.2.1 Explain why $x^2 + 10x$ can be represented by the L-shaped region shown in the left half of Figure 2.6.

2.2.2 Deduce that the black square that completes the L-shaped region to a big square has area 25, shown in the right half of Figure 2.6.

2.2.3 Conclude that the big square has area 64, and hence that $x = 3$ (why?) is a solution of $x^2 + 10x = 39$.

2.2.4 Find *another* solution of $x^2 + 10x = 39$. Why does al-Khwārizmī's method not discover it?

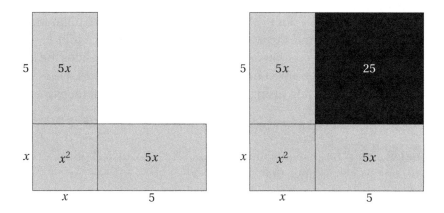

Figure 2.6: Completing the square.

The examples above point to a limitation of al-Khwārizmī's method: it does not necessarily find all solutions of a quadratic equation. To get a clearer picture of the solutions, it helps to use the discriminant.

2.2.5 By calculating the discriminant, show that the equation

$$x^2 + x + 1 = 0$$

has no real solutions.

2.2.6 Explain, with the help of the discriminant, why the equation

$$x^2 + 6x + 9 = 0$$

has exactly one solution. Also, find the solution.

2.3 Solving Cubic Equations

To see why Cardano was so excited about solving cubic equations, it helps to try solving one yourself. Take the equation $x^3 + 6x - 4 = 0$, for example. I think you will soon agree that, when you know only the solution of quadratic equations, it is hard to see where to start. Yet there is a simple trick that solves $x^3 + 6x - 4 = 0$, by solving *only* a quadratic equation, and taking cube roots of its solutions. Here is how it goes.

The trick is to let $x = u + v$. We now have to solve for two unknowns, u and v, instead of one, but we have an extra degree of freedom.

We rewrite the equation as $x^3 = -6x + 4$ and first consider the left side, x^3. Since $x = u + v$ we find x^3 by multiplying $(u + v)(u + v)(u + v)$, which gives the left side (using the distributive law several times)

$$x^3 = u^3 + 3u^2v + 3uv^2 + v^3 = 3uv(u + v) + u^3 + v^3.$$

The left side x^3 has to equal the right side, $-6x + 4$, which is $-6(u+v)+4$. Now one way to arrange that

$$3uv(u + v) + u^3 + v^3 = -6(u + v) + 4$$

is to find u and v such that

$$3uv = -6 \quad \text{and} \quad u^3 + v^3 = 4. \tag{1}$$

This is possible, thanks to our freedom to choose u as we please.

From the first of the equations (1) we get $v = -2/u$, and substituting this in the second equation of (1) gives

$$u^3 - \frac{2^3}{u^3} = 4, \quad \text{that is,} \quad u^3 - \frac{8}{u^3} = 4.$$

Multiplying both sides of this by u^3 we get

$$\left(u^3\right)^2 - 8 = 4u^3 \quad \text{or} \quad \left(u^3\right)^2 - 4u^3 - 8 = 0,$$

which is a quadratic equation in u^3. We solve for u^3 by the quadratic formula:

$$\begin{aligned} u^3 &= \frac{4 \pm \sqrt{4^2 + 4 \times 8}}{2} \\ &= \frac{4 \pm \sqrt{48}}{2} \\ &= \frac{4 \pm 4\sqrt{3}}{2} \\ &= 2 \pm 2\sqrt{3}. \end{aligned}$$

Now looking back at the two equations (1), we see that u and v are interchangeable, so v^3 satisfies the same quadratic equation as u^3. We can choose either of the values $2 + 2\sqrt{3}$, $2 - 2\sqrt{3}$ to be u^3, and then the other will be v^3, because $u^3 + v^3 = 4$. Consequently, whichever of the values $\sqrt[3]{2 + 2\sqrt{3}}$, $\sqrt[3]{2 - 2\sqrt{3}}$ we take for u, the other will be v, and therefore

$$x = u + v = \sqrt[3]{2 + 2\sqrt{3}} + \sqrt[3]{2 - 2\sqrt{3}}.$$

The brilliant trick of setting x equal to $u + v$ works for any equation of the form $x^3 = px + q$, and it leads to the so-called *Cardano formula*

$$x = \sqrt[3]{\frac{q}{2} + \sqrt{\left(\frac{q}{2}\right)^2 - \left(\frac{p}{3}\right)^3}} + \sqrt[3]{\frac{q}{2} - \sqrt{\left(\frac{q}{2}\right)^2 - \left(\frac{p}{3}\right)^3}}.$$

But is this always a solution? Could it be, as with quadratic equations, that the formula signals NO SOLUTION when square roots of negative numbers occur? For reasons that will be explained in the next section, the acid test of the formula is the equation

$$x^3 = 15x + 4.$$

For this equation, $p/3 = 5$ and $q/2 = 2$, so the Cardano formula gives

$$x = \sqrt[3]{2 + \sqrt{2^2 - 5^3}} + \sqrt[3]{2 - \sqrt{2^2 - 5^3}}$$

$$= \sqrt[3]{2 + \sqrt{-121}} + \sqrt[3]{2 - \sqrt{-121}}$$

$$= \sqrt[3]{2 + 11\sqrt{-1}} + \sqrt[3]{2 - 11\sqrt{-1}}.$$

With the occurrences of $\sqrt{-1}$, this certainly looks like "imaginary solution." But it shouldn't, because *the equation* $x^3 = 15x + 4$ *has an obvious solution!* Namely, $x = 4$, because $4^3 = 64 = 15 \times 4 + 4$.

Can it really be that $\sqrt[3]{2 + 11\sqrt{-1}} + \sqrt[3]{2 - 11\sqrt{-1}} = 4$?

Exercises

2.3.1 Find a real solution of $x^3 = 6x - 4$ by trial and error.

2.3.2 Write down the solution of $x^3 = 6x - 4$ given by the Cardano formula, and show that it simplifies to

$$x = \sqrt[3]{-2 + 2\sqrt{-1}} + \sqrt[3]{-2 - 2\sqrt{-1}}$$

2.4 Real Solutions via Imaginary Numbers

The two apparently different solutions of the equation $x^3 = 15x + 4$, namely $x = 4$ and $x = \sqrt[3]{2 + 11\sqrt{-1}} + \sqrt[3]{2 - 11\sqrt{-1}}$, were first noticed in 1572 by the Italian mathematician Rafael Bombelli. He drew attention to them in his book *Algebra*, and cleverly reconciled them by showing that

$$2 + 11\sqrt{-1} = (2 + \sqrt{-1})^3, \quad \text{hence} \quad \sqrt[3]{2 + 11\sqrt{-1}} = 2 + \sqrt{-1},$$

$$2 - 11\sqrt{-1} = (2 - \sqrt{-1})^3, \quad \text{hence} \quad \sqrt[3]{2 - 11\sqrt{-1}} = 2 - \sqrt{-1},$$

and therefore

$$\sqrt[3]{2 + 11\sqrt{-1}} + \sqrt[3]{2 - 11\sqrt{-1}} = 2 + \sqrt{-1} + 2 - \sqrt{-1}$$
$$= 4.$$

Guessing that $\sqrt[3]{2 + 11\sqrt{-1}} = 2 + \sqrt{-1}$ and $\sqrt[3]{2 - 11\sqrt{-1}} = 2 - \sqrt{-1}$, so $\sqrt{-1}$ and $-\sqrt{-1}$ cancel in the sum of cube roots, was a brilliant move. But Bombelli's most important step was to *assume that $\sqrt{-1}$ obeys the ordinary rules of algebra.* In particular, he assumed that

$$(\sqrt{-1})^2 = -1 \quad \text{and hence} \quad (\sqrt{-1})^3 = -1\sqrt{-1} = -\sqrt{-1}.$$

Along with other algebraic rules, such as the distributive law, this makes it possible to calculate

$$(2 + \sqrt{-1})^3 = 2^3 + 3 \cdot 2^2 \sqrt{-1} + 3 \cdot 2(\sqrt{-1})^2 + (\sqrt{-1})^3$$
$$= 8 + 12\sqrt{-1} - 6 - \sqrt{-1}$$
$$= 2 + 11\sqrt{-1}$$

—the result claimed by Bombelli. A similar calculation explains his claim that

$$(2 - \sqrt{-1})^3 = 2 - 11\sqrt{-1},$$

as required for the miraculous cancellation in the sum of cube roots.

Thus reconciliation of the obvious solution of $x^3 = 15x + 4$ with the formula solution was found by extending the rule of law from real arithmetic to arithmetic with $\sqrt{-1}$. This was the first indication that the "imaginary number" $\sqrt{-1}$ is a meaningful and useful idea; in fact a *necessary* idea if one is to operate with cubic equations.

Exercises

2.4.1 Use the formula for $(a + b)^3$ in Exercise 2.1.2 to show that

$$(1 + \sqrt{-1})^3 = -2 + 2\sqrt{-1} \quad \text{and} \quad (1 - \sqrt{-1})^3 = -2 - 2\sqrt{-1}.$$

2.4.2 Use the results of 2.4.1 to show that

$$\sqrt[3]{-2 + 2\sqrt{-1}} + \sqrt[3]{-2 - 2\sqrt{-1}} = 2$$

(which reconciles the solutions of $x^3 = 6x - 4$ found in Exercises 2.3.1 and 2.3.2).

2.5 Where Were Imaginary Numbers before 1572?

Many mathematicians believe they *discover* mathematics, rather than invent it, just as astronomers discover stars and planets or chemists discover elements. This parallel between mathematics and science has many aspects, some of which we return to later in the book. For the moment I want to illustrate the aspect of *unrecognized appearances*.

One of the most famous events in the history of astronomy was the discovery of the planet Neptune by Adams and Leverrier in 1846. This "discovery" was really when Neptune was first recognized as a planet, because telescopes had swept the skies for more than two centuries before 1846, and undoubtedly somebody saw Neptune but took it for just another star. In fact, we now know Neptune's motion well enough to calculate its position at any time over the last few centuries, and it has been found that *Galileo saw Neptune in 1612, without recognizing that it was a planet!* His notes for 28 December 1612 and 2 January 1613 record what he believed to be a star, but in the position then occupied by Neptune. (Galileo even noticed that the "star" appeared to have moved between his two observations, but put this down to experimental error. See Kowal and Drake, Galileo's Observations of Neptune, *Nature*, 28, 311 (1980).)

Now if "imaginary" numbers are a significant concept, and not just a device to fix the Cardano formula, their effects should have been felt in mathematics before 1572. If $\sqrt{-1}$ lives among the numbers, there is an arithmetic of *complex numbers* $a + b\sqrt{-1}$ or $a + bi$, where a and b are real and $i^2 = -1$. This arithmetic indeed produces observable effects.

The *sum* of complex numbers is very straightforward:

$$(a + bi) + (c + di) = (a + c) + (b + d)i.$$

It is simply the result of adding the "real parts" a, c and the "imaginary parts" b, d separately. However, when we form the *product* of complex numbers, the real and imaginary parts interact in an interesting way. By using the distributive law and the rule $i^2 = -1$, we find

$$(a + bi)(c + di) = ac + bdi^2 + bci + adi$$
$$= (ac - bd) + (bc + ad)i.$$

Thus the product of the number with real and imaginary parts a, b and the number with real and imaginary parts c, d is the number with real and imaginary parts $ac - bd$, $bc + ad$.

This interaction between pairs of real numbers was observed long before 1572! The first "sighting" is probably in the work of Diophantus around 200 CE. In his *Arithmetica*, [25, Book III, Problem 19], he makes the following cryptic remark:

> 65 is naturally divided into two squares in two ways, namely $7^2 + 4^2$ and $8^2 + 1^2$, which is due to the fact that 65 is the product of 13 and 5, each of which is the sum of two squares.

He seems to know that the product of sums of two squares is itself a sum of two squares, and specifically that

$$(a^2 + b^2)(c^2 + d^2) = (ac - bd)^2 + (bc + ad)^2. \tag{2}$$

His splitting of 65 into $7^2 + 4^2$ comes from $a = 3$, $b = 2$, $c = 2$, $d = 1$, namely

$$65 = 13 \times 5 = (3^2 + 2^2)(2^2 + 1^2) = (3 \times 2 - 2 \times 1)^2 + (2 \times 2 + 3 \times 1)^2 = 4^2 + 7^2.$$

His second splitting, $8^2 + 1^2$, comes from the variant identity

$$(a^2 + b^2)(c^2 + d^2) = (ac + bd)^2 + (bc - ad)^2.$$

Admittedly, Diophantus gives only a single instance of (2), but this is his style. His notation does not have symbols for more than one variable, so he expects his readers to infer the general rule from well-chosen examples. The general formula (2) was recognized explicitly by Abū Ja'far al-Khazin around 950 CE, commenting on this problem of Diophantus, and proved by Fibonacci in his *Book of Squares* of 1225. For Fibonacci's very laborious proof, see [19, pp. 23–28].

For Diophantus, (2) gives a *rule* for splitting a product of sums of squares into a sum of squares: if two numbers split into the squares of a, b and c, d, then their product splits into the squares of $ac - bd$, $bc + ad$. Thus the rule of Diophantus has precisely the same form as the rule for multiplying complex numbers. Is this just a coincidence? No, because there is more …

Diophantus is interested in sums of squares, but he views $a^2 + b^2$ as the *square on the hypotenuse* of a right-angled triangle with perpendicular sides a and b. Thus he definitely associates $a^2 + b^2$ with a *pair* a, b. The rule behind (2) takes two right-angled triangles and finds a third (which we might call the "product" triangle) whose hypotenuse is the product of the hypotenuses of the first two (Figure 2.7).

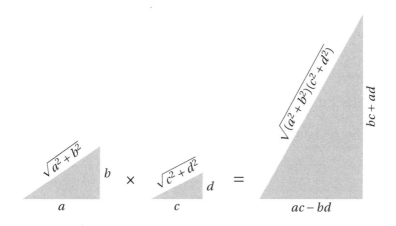

Figure 2.7: The Diophantus "product" of triangles.

This picture is extraordinarily close to what we now regard as the "right" way to interpret complex numbers. The number $a + bi$ is viewed as the point in the plane at horizontal distance a, and vertical distance

b, from the zero point or *origin* O. Thus $a + bi$ is the top right corner of the triangle with bottom left corner at O. The length $\sqrt{a^2 + b^2}$ of the hypotenuse is called the *absolute value*, $|a + bi|$, of $a + bi$, and (2) expresses the so-called *multiplicative property of absolute value*:

$$|a + bi||c + di| = |(a + bi)(c + di)|. \tag{3}$$

To be precise, (2) is what we get by squaring equation (3), because

$$|a + bi|^2 = a^2 + b^2,$$
$$|c + di|^2 = c^2 + d^2,$$
$$|(a + bi)(c + di)|^2 = |(ac - bd) + (bc + ad)i| = (ac - bd)^2 + (bc + ad)^2.$$

Another good thing about equation (2), as a starting point for the study of complex numbers, is that it involves nothing "imaginary." The equation is simply a property of real numbers, and it can be verified by multiplying out both sides. It begins to look as though finding complex numbers through the square root of -1 is a lucky accident. And yet ... equation (2) is kind of an accident too. Who would have guessed in advance that the product of sums of two squares is a sum of two squares? Is this true for three squares? Four? Five? See Chapter 6 for more on these questions.

Exercises

2.5.1 Check that $533 = (5^2 + 4^2)(3^2 + 2^2)$.

2.5.2 Use 2.5.1 and the formula

$$(a^2 + b^2)(c^2 + d^2) = (ac - bd)^2 + (bc + ad)^2$$

to express the number 533 as a sum of two squares.

2.5.3 Calculate $(5 + 4i)(3 + 2i)$. How is the result related to the answer to Exercise 2.5.2?

2.5.4 Check that $1802 = (5^2 + 3^2)(7^2 + 2^2)$.

2.5.5 Find two squares with sum 1802 by calculating $(5 + 3i)(7 + 2i)$.

2.6 Geometry of Multiplication

Complex numbers appeared sporadically in mathematics for more than 200 years after Bombelli's *Algebra*, but without being fully accepted or recognized. Sometimes it was a case of using $\sqrt{-1}$ to obtain a result in algebra that would later be explained by an alternative, purely real, calculation. At other times, mathematicians stumbled on unexpected properties of real numbers that we now prefer to explain as effects of $\sqrt{-1}$.

One such effect was discovered by the French mathematician Francois Viète around 1590. In his *Genesis triangulorum*, Viète studied the Diophantus "product" of triangles and found a second feature just as remarkable as the multiplicative property of hypotenuses—an *additive property of angles*. If the angles of the two given triangles are θ and φ, as shown in Figure 2.8, then the angle of the "product" triangle is $\theta + \varphi$.

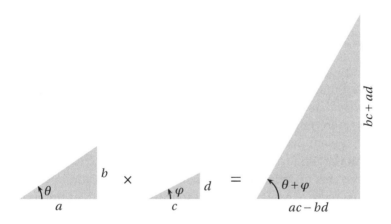

Figure 2.8: Additivity of angles.

Compare this with the situation found in Section 1.1. There the audible quantity called "pitch" was added when frequency was multiplied. Here we have a visible quantity called "angle" which is added when triangles (or, at least, their hypotenuses) are multiplied. Another example of multiplication perceived as addition! Is this "multiplication" the same as the previous one?

Around 1800 the results of Diophantus and Viète were seen to be parts of the same simple picture: a *number plane*, in which the complex number $a + bi$ is represented by the point (a, b) at horizontal distance a and vertical distance b from the origin O (Figure 2.9, with arrows indicating the positive directions horizontally and vertically).

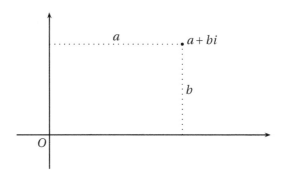

Figure 2.9: Complex numbers as points in the plane.

This picture is in the background of several eighteenth-century applications of complex numbers. It was brought into the foreground by the Norwegian surveyor Caspar Wessel in 1797. Wessel's paper went unnoticed for nearly 100 years, but the same idea was put forward by others. It went mainstream in 1832, when the great German mathematician Carl Friedrich Gauss used the geometry of complex numbers to study sums of squares. We pick up this thread again in Chapter 7. Here our goal is to explain how multiplication of complex numbers can be perceived as addition of angles.

The most important feature of the number plane is that *length or distance represents absolute value*. The Pythagorean theorem says that the distance from O to $a + bi$ is $\sqrt{a^2 + b^2} = |a + bi|$, and it follows easily that the distance between any two complex numbers is the absolute value of their difference.

Now suppose that u is some complex number and that we multiply all the numbers in the plane by u. This sends numbers v and w to the numbers uv and uw. The distance between uv and uw is $|uw - uv|$, and

$$|uw - uv| = |u(w - v)| \quad \text{by the distributive law}$$
$$= |u||w - v| \quad \text{by the multiplicative property}$$
$$= |u| \times \text{distance between } v \text{ and } w.$$

Thus all distances in the plane are multiplied by $|u|$. Multiplying the number plane by a constant number u therefore induces a *magnification by $|u|$*—just as it does on the number line.

But multiplication by u also induces a *rotation* of the plane, and we see this most clearly when the magnification factor $|u|$ is 1. In this case, multiplication by u leaves all distances unchanged, so the plane is moved rigidly. Also the point O remains fixed, because $0 \times u = 0$. And if $u \neq 1$, then O is the *only* fixed point, because in that case $uv \neq v$ if $v \neq 0$. Hence *multiplication by a complex number u with absolute value 1 is a rotation about O.*

This rotation necessarily sends the point 1 to $u \times 1 = u$, hence the rotation is necessarily through the angle θ between the direction of u from O and the real axis (Figure 2.10).

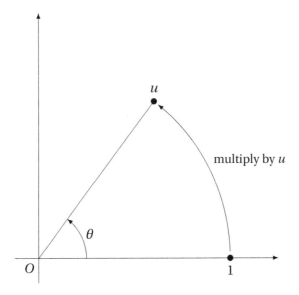

Figure 2.10: The angle of rotation is the angle of the multiplier.

Now suppose that u and v are two complex numbers of absolute value 1, with angles θ and φ, respectively. We can multiply by uv by first multiplying by u, which rotates the plane through angle θ, then multiplying by v, which rotates the plane through angle φ. *The angle of the product uv is therefore $\theta + \varphi$—the sum of the angles of u and v.* This explains the discovery of Viète about addition of angles, and shows that the "multiplication" in question is truly multiplication of numbers.

To sum up: *the complex number system is an extension of the real number system in which multiplication is "audible," in one sense, and "visible" in another sense.* If we factor a complex number u into the real number $|u|$ and the complex number $u/|u|$, then multiplication by u is a combination of multiplication by $|u|$ and multiplication by $u/|u|$.

- Multiplication by $|u|$ is the "audible" part, since multiplying frequency by a real constant can be perceived as a constant addition to pitch.

- Multiplication by $u/|u|$ is the "visible" part, since $u/|u|$ has absolute value 1, so multiplying by it can be perceived as a rotation of the number plane.

Rotation and Complex nth Roots of 1

Multiplication by the number i rotates the number plane through a right angle, or quarter turn, as is clear from Figure 2.11. This confirms that $i^2 = -1$, because two quarter turns make a half turn, which is the same as multiplication by -1. Multiplication by $\sqrt{-1}$ is therefore a "square root of a half turn," that is, a rotation through a right angle.

We could also call multiplication by $\sqrt{-1}$ a "fourth root of a full turn," since $(\sqrt{-1})^4 = 1$. Indeed, for $n = 2, 3, 4, \ldots$ there is a complex number, usually denoted by ζ_n ("zeta sub n"), such that multiplication by ζ_n rotates the plane through $1/n$ of a full turn. ζ_n is simply the result rotating the point 1 through $1/n$ of a full turn, and it is called an nth *root of* 1 because $\zeta_n^n = 1$. It is easy to see that all its powers, $\zeta_n^2, \zeta_n^3, \ldots$ are also roots of 1.

A remarkable fact about roots of 1 is that only the square roots ± 1 and the fourth roots $\pm i$ are "rational" complex numbers, that is, numbers of the form $\frac{a}{c} + \frac{b}{c}i$ for natural numbers a, b, and c. We prove this

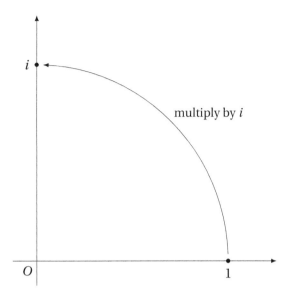

Figure 2.11: Multiplication by i.

in Section 7.6, through a deeper study of complex numbers. It follows that a "rational" right-angled triangle—one with natural number sides a, b, c such that $a^2 + b^2 = c^2$—does not have a "rational" angle (m/n of a full turn for natural numbers m and n). This explains something about the sequence of angles shown in Figure 1.5 (the ones derived from the Pythagorean triples of Plimpton 322). The sequence is necessarily irregular, because a right angle cannot be divided into equal parts by right-angled triangles with natural number sides.

It also establishes a rather nice parallel between musical scales and scales of angles, showing that the rational/irrational distinction is just as interesting for complex numbers as it is for the reals.

- Equal pitch intervals are obtained by multiplying frequency by a constant ratio. It is impossible to divide the natural interval, the octave, into equal parts by rational frequency ratios.

- Likewise, equal angles are obtained by multiplying by a constant complex number of absolute value 1. It is impossible to divide the natural angle, the right angle, into equal parts by the hypotenuses of rational right-angled triangles.

Exercises

Let θ be the angle that the line from O to the point $3 + 4i$ makes with the x-axis (Figure 2.12).

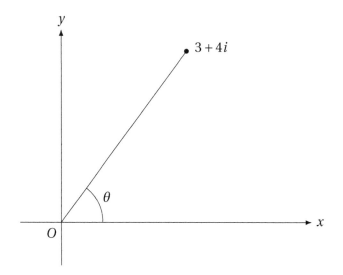

Figure 2.12: Geometry of the point $3 + 4i$.

2.6.1 What is the angle from O to the point $(3 + 4i)^2$?

2.6.2 Explain why the line from O to the point $2 + i$ makes angle $\theta/2$ with x-axis.

The complex number $\zeta_3 = z$ rotates the plane through $1/3$ of a turn and hence the numbers z, z^2, and $z^3 = 1$ are positioned at the corners of an equilateral triangle.

2.6.3 Explain why the equation $z^3 = 1$ satisfied by z is equivalent to the equation $(z - 1)(z^2 + z + 1) = 0$

2.6.4 Deduce that the three cube roots of 1 are 1, $\frac{-1 + i\sqrt{3}}{2}$, and $\frac{-1 - i\sqrt{3}}{2}$.

This result confirms, since we know that $\sqrt{3}$ is irrational by Exercise 1.3.5, that the cube roots of 1 (other than 1 itself) are irrational complex numbers.

2.6.5 Draw a picture of the equilateral triangle formed by the three cube roots of 1, and use it to find three more points which, together with the corners of the triangle, are the corners of a regular hexagon.

2.7 Complex Numbers Give More than We Asked for

It has been written that the shortest and best way between two truths of the real domain often passes through the imaginary one.

Jacques Hadamard, *An Essay on the Psychology of Invention in the Mathematical Field*, Chapter VIII

Accepting the number $\sqrt{-1}$ to solve the cubic equation $x^3 = 15x + 4$, led mathematicians to think again about quadratic equations such as $x^2 = -1$. In the realm of complex numbers, the "imaginary" solutions $x = \pm i$ really exist, and so do the solutions

$$x = \frac{-b \pm \sqrt{b^2 - 4ac}}{2a}$$

of the general quadratic equation $ax^2 + bx + c = 0$.

As we have seen in Section 1.3, any cubic equation of the form $x^3 = px + q$ can be solved by solving a quadratic equation and taking cube roots. An arbitrary cubic equation $ax^3 + bx^2 + cx + d = 0$ can be reduced to the form $x^3 = px + q$, so in fact all cubic equations may be solved by reducing them to quadratics. Cardano's *Ars Magna* of 1545 has methods for solving all types of cubics—and more. It also contains the solution of *quartic* equations (those in which the highest power of x is x^4), discovered by Cardano's student Lodovico Ferrari. The general quartic equation is solved via cubic and quadratic equations, so it too has solutions in the complex numbers.

Thus the solution $x = i$ of the special equation $x^2 = -1$ brings with it the solutions of *all* equations of degree ≤ 4. Following this unbroken run of algebraic success, a more ambitious question began to take shape: does every equation of degree n have n solutions in the complex numbers? The French mathematician Albert Girard was the first to raise this possibility, writing in his *L'Invention nouvelle in l'algebra* of 1629 that

> Every equation of algebra has as many solutions as the exponent of the highest term indicates.

Girard's conjecture is now known as the *fundamental theorem of algebra*. It was echoed by his compatriot René Descartes in his *Geometry* of 1637, with the support of Descartes' result now known as the *factor theorem*: if $x = x_1$ is a solution of the equation $p(x) = 0$, then $(x - x_1)$ is a factor of $p(x)$, that is

$$p(x) = (x - x_1)q(x).$$

Here the degree of $q(x)$ (the highest power of x in $q(x)$) is necessarily one less than the degree of $p(x)$. If $q(x) = 0$ has a solution $x = x_2$, it likewise follows that

$$q(x) = (x - x_2)r(x), \quad \text{and hence} \quad p(x) = (x - x_1)(x - x_2)r(x),$$

where the degree of $r(x)$ is one less than the degree of $q(x)$, and so on. Thus if each equation has a solution, an equation $p(x) = 0$ of degree n in fact has n solutions, $x = x_1, x = x_2, \ldots, x = x_n$.

The factor theorem is easy to prove, so the hard part of the fundamental theorem of algebra is proving the existence of *one* solution for each equation $p(x) = 0$. The quadratic, cubic, and quartic equations are misleading in a way, because a formidable obstacle occurs at the next level. When $p(x)$ has degree 5 the equation $p(x) = 0$ *cannot* in general be reduced to equations of lower degree. This delayed the proof of the fundamental theorem of algebra for a long time, while futile attempts were made to solve the equation of degree 5 via equations of lower degree.

Finally, in 1799, Gauss tried a new approach: he set out to prove only the *existence* of solutions, rather than trying to construct them by square roots, cube roots, and so on. Gauss's first attempt had a serious gap, but the general idea was sound, and he later presented satisfactory proofs. Most of them use the geometry of complex numbers in an essential way, though one of them (Gauss's second proof, 1816) reduces the role of complex numbers to an elegant minimum, using them only to solve quadratic equations. He shows that solving an equation of any degree may be reduced to

- solving quadratic equations, and

- solving an arbitrary equation of *odd* degree, that is, an equation in which the highest power of x is odd.

An odd-degree equation, say

$$x^{2m+1} + \text{terms involving lower powers of } x = 0$$

always has a solution because the values of the left side run continuously from large negative values (for large negative x) to large positive values (for large positive x), hence for some x the value of the left side is zero. This is perhaps the best way to explain why solving $x^2 = -1$ paves the way for solving equations of any degree.

However, this barely begins to describe what $\sqrt{-1}$ can do. So far we have considered the role of $\sqrt{-1}$ only in *polynomial* functions

$$p(x) = a_n x^n + a_{n-1} x^{n-1} + \cdots + a_1 x + a_0.$$

More surprises occur when complex numbers are input to other functions, such as $\cos x$, $\sin x$, and e^x. One meets the cosine and sine functions in geometry, where $\cos\theta$ and $\sin\theta$ are the width and height of a right-angled triangle with hypotenuse 1 and angle θ (Figure 2.13).

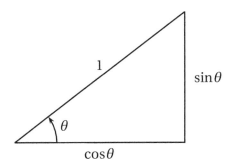

Figure 2.13: The cosine and sine functions.

This implies that $\cos\theta + i\sin\theta$ is a complex number of absolute value 1. Hence multiplying by $\cos\theta + i\sin\theta$ rotates the number plane through angle θ, by the geometric properties of multiplication found in the previous section.

The exponential function e^x is harder to explain: e is a certain number, approximately 2.718 and, as the real numbers x fill the spaces between 1, 2, 3, ..., the values of e^x fill the spaces between $e, e^2, e^3, \ldots$. When x is real, there seems to be no connection between e^x and $\cos x$ and $\sin x$. But in the world of complex numbers, there is an amazing connection, discovered by the Swiss mathematician Leonhard Euler in 1748:

$$e^{ix} = \cos x + i \sin x. \tag{4}$$

This incredible result may be found in Euler's book, *Introduction to the Analysis of the Infinite*. The result (4) becomes credible if one knows that the function e^x turns sums into products,

$$e^{u+v} = e^u e^v,$$

because the function $\cos x + i \sin x$ does the same thing:

$$\cos(u + v) + i \sin(u + v) = (\cos u + i \sin u)(\cos v + i \sin v), \tag{5}$$

since the left side of (5) rotates the plane through $u + v$, and the right side rotates the plane through u, then v.

The connection between $\cos x$, $\sin x$, and e^x revealed by complex numbers has countless applications, from engineering and physics to the almost mystical equation that results from putting $x = \pi$ in (4):

$$e^{i\pi} = -1.$$

Alas, we do not have enough space to explain this, but you can learn more from two excellent books: *An Imaginary Tale* by Paul Nahin, and *Visual Complex Analysis* by Tristan Needham.

Exercises

The formula (5) above enables us to prove the so-called *addition formulas* for sine and cosine, without assuming their connection with the exponential function.

2.7.1 By expanding the right side of (5), then comparing it with the left side, prove that

$$\cos(u + v) = \cos u \cos v - \sin u \sin v,$$
$$\sin(u + v) = \sin u \cos v + \cos u \sin v.$$

2.7.2 Deduce from these formulas that

$$\tan(u+v) = \frac{\tan u + \tan v}{1 - \tan u \tan v}.$$

With the formula for $\tan(u+v)$ we can confirm the additive angle property for "triangle multiplication" discovered by Viète and illustrated by Figure 2.8.

2.7.3 Given that $\tan\theta = \frac{b}{a}$ and $\tan\varphi = \frac{d}{c}$, prove that

$$\tan(\theta+\varphi) = \frac{bc+ad}{ac-bd}.$$

2.8   Why Call Them "Complex" Numbers?

The word "complex" was introduced in a well-meaning attempt to dispel the mystery surrounding "imaginary" or "impossible" numbers, and (presumably) because two dimensions are more complex than one. Today, "complex" no longer seems such a good choice of word. It is usually interpreted as "complicated," and hence is almost as prejudicial as its predecessors. Why frighten people unnecessarily? If you are not sure what "analysis" is, you won't want to know about "complex analysis"—but it is the best part of analysis!

In fact, complex numbers are not much more complicated than reals, and many *structures* built on the complex numbers actually have simpler behavior than the corresponding structures built on the real numbers.

Polynomials are one example. We have just seen that a polynomial of degree n always splits into n complex factors $x - x_1$, $x - x_2$, ..., $x - x_n$, corresponding to the n solutions $x = x_1, x_2, \ldots, x_n$ of the equation $p(x) = 0$. (In fact, we say there are n solutions *because* there are n factors; it doesn't matter if some of $x_1, x_2, \ldots, x_n$ are equal.) Splitting a polynomial into real factors is an entirely different proposition. Even when the degree is 2, we sometimes have two factors, as with $x^2+2x+1 = (x+1)(x+1)$, and sometimes only one, as with x^2+1, which does not split any further.

A second example, which builds on factorization, is counting the intersections of curves. We define an *algebraic curve* to be one whose points are pairs (x, y) satisfying a polynomial equation in two variables $p(x, y) = 0$. Initially, we interpret x as the horizontal distance from the origin and y as the vertical distance. For example, the equation

$$x^2 + y^2 = 1 \quad \text{or} \quad x^2 + y^2 - 1 = 0$$

defines the points at distance 1 from the origin in the plane, which form the circle of radius 1 with center O. The circle is therefore an algebraic curve, and we say it has *degree 2* because the highest power in its equation is 2. Likewise,

$$x = y \quad \text{or} \quad x - y = 0$$

defines an algebraic curve of degree 1, and it is clearly a line through O. The line meets the circle in two points, so the number of intersection points in this case happens to be the product of the degrees.

From a few such examples, Isaac Newton [42, vol. 1, p. 498] in 1665 launched the following bold speculation (he uses the term "dimension" for degree, "lines" for curves, and "rectangle" for product):

> For ye number of points in wch two lines may intersect can never bee greater y^n ye rectangles of ye numbers of their dimensions. And they always intersect in soe many points, excepting those wch are imaginarie onely.

The modern version of Newton's claim is called *Bézout's theorem*, and it says that a curve of degree m meets a curve of degree n in mn points. There are several conditions involved in counting points correctly, but the most important is that *complex points must be allowed*. Under reasonable conditions, the problem of finding the intersections of a curve of degree m with a curve of degree n can be reduced to solving an equation of degree mn, and to get mn solutions it is essential, as we know, to allow solutions to be complex. Thus we have to accept "curves" including complex points, that is, pairs (x, y) of complex numbers x and y. But it's worth it to get Bézout's theorem.

The behavior of real algebraic curves is horribly complicated in comparison, with almost no connection between degree and the number of intersections. For example, a circle (degree 2) can meet a straight line (degree 1) in two real points, which may be equal, or none. Bézout's theorem truly shows that complex curves are "simple," and real curves are "complex."

Likewise, complex functions are actually better behaved than real functions, and the subject of complex analysis is known for its regularity and order, while real analysis is known for wildness and pathology. A smooth complex function is *predictable*, in the sense that the values of the function in an arbitrarily small region determine its values everywhere. A smooth real function can be completely unpredictable: for example, it can be constantly zero for a long interval, then smoothly change to the value 1.

The worst aspect of the term "complex"—one that condemns it to eventual extinction in my opinion—is that it is also applied to structures called "simple." Mathematics uses the word "simple" as a technical term for objects that cannot be "simplified." Prime numbers are the kind of thing that might be called "simple" (though in their case it is not usually done) because they cannot be written as products of smaller numbers. At any rate, some of the "simple" structures are built on the complex numbers, so mathematicians are obliged to speak of such things as "complex simple Lie groups." This is an embarrassment in a subject that prides itself on consistency, and surely either the word "simple" or the word "complex" has to go.

There is also the possibility that there are numbers "more complex" than the complex numbers. What are you going to call them? Hypercomplex? Good guess. There are such things, and we shall meet them (and a better name for them) in Chapter 6.

Chapter 3

The Horizon

Preview

In the fifteenth century, artists discovered *perspective*, which revolutionized the drawing of three-dimensional scenes (compare Figures 3.1[1] and 3.2). The discovery also revolutionized geometry with a new *geometry of vision*, quite unlike the old geometry of measurement.

Figure 3.1: Picture from before the discovery of perspective.

---

[1] The original of Figure 3.1 is in a handpainted manuscript in the British Library, *The Lives of Sts. Edmund and Fremund* by John Lydgate, from around 1434. The version here is an accurate but sharper nineteenth-century copy by Henry Ward, in the Wellcome library.

Figure 3.2: After the discovery of perspective: Albrecht Dürer's *Saint Jerome*.

However, making sense of the geometry of vision is harder than it looks, because the eye "sees" points that do not exist. Mathematicians called them "ideal" or "imaginary" points before settling on the term *points at infinity.*

The points at infinity form a line, the horizon, *where impossible things happen, such as meetings of parallel lines.* How geometry absorbs this paradox, and thereby becomes bigger and better, is the subject of this chapter.

3.1 Parallel Lines

As I understand it, the first time Gabriel García Márquez opened Kafka's *The Metamorphosis*, he was a teenager, reclining on a couch. Upon reading

> As Gregor Samsa awoke one morning from uneasy dreams he found himself transformed in his bed into a gigantic insect ...

García Márquez fell off his couch, astonished by the revelation that you were *allowed* to write like that!

... It has happened to me often, and surely a similar thing happens to all mathematicians, that upon hearing of someone's new idea, or new construction, I have, like García Márquez, fallen off my (figurative) couch, thinking in amazement, "I didn't realize we were *allowed* to do that!"

Barry Mazur, *Imagining Numbers*, Penguin 2003, pp. 64 and 69

One of the key concepts in geometry is that of *parallel lines*, which can be simply described as lines in the plane that do not meet. The purpose of this chapter is to explain why parallel lines are important, and to show that *they become even more important when allowed to meet!* This paradox—of allowing lines to meet when they don't—will be explained in Section 3.3. In the present section we set the stage by reviewing the basic properties of parallels and some of their consequences.

First, parallels exist and they are unique. A precise statement of this property is the following:

Parallel axiom. *If $\mathcal{L}$ is any line and P is a point outside it, then there is exactly one line through P that does not meet $\mathcal{L}$.*

This unique line is called the *parallel to $\mathcal{L}$ through P* (Figure 3.3).

In calling the existence and uniqueness of parallels an "axiom" we are not suggesting there is any doubt about its truth. Rather, we are stressing its role as a *starting point*, from which other truths follow. The role of parallel lines in geometry was first recognized by Euclid, who

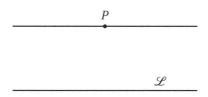

Figure 3.3: Parallel lines.

demonstrated in his *Elements* that many theorems of geometry follow from the existence and uniqueness of parallels.

Euclid did not state the parallel axiom as we have; in fact his version is much more cumbersome. The Greeks did not accept infinite lines, only the possibility of extending a finite line indefinitely, so Euclid had to state the parallel axiom as a condition under which finite lines *do* meet, if extended sufficiently far. His statement is equivalent to the parallel axiom above, but goes as follows.

Euclid's parallel axiom. *If a line $\mathcal{N}$, crossing lines $\mathcal{L}$ and $\mathcal{M}$, makes angles α and β with them on the same side with $\alpha + \beta$ less than two right angles, then $\mathcal{L}$ and $\mathcal{M}$ meet on this side if extended sufficiently far.* (Figure 3.4).

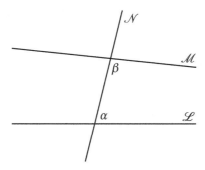

Figure 3.4: Non-parallel lines.

Notice that, in order to state his version of the axiom, Euclid needs an auxiliary line and also the seemingly irrelevant concept of angle. It is striking how much simpler the parallel axiom becomes when lines are assumed to be infinite. Nevertheless, Euclid's parallel axiom also has its merits. It suggests that parallels control the behavior of angles and, presumably, the behavior of lengths and areas as well. This is indeed the case. Here are just a few of the consequences of the parallel axiom that may be found in the *Elements*:

- Rectangles exist.

- The Pythagorean theorem.

- The angles of any triangle sum to two right angles.

Exercises

A basic fact about parallel lines $\mathscr{L}$ and $\mathscr{M}$ is that, when a line $\mathscr{N}$ crosses them, equal angles α are formed in the positions shown in Figure 3.5.

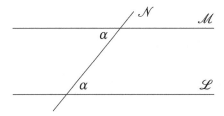

Figure 3.5: Parallel lines and equal angles.

3.1.1 Use the parallel lines $\mathscr{L}$ and $\mathscr{M}$ to find some equal angles in Figure 3.6.

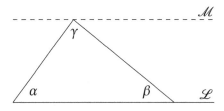

Figure 3.6: Angle sum of a triangle.

Hence show that $\alpha + \beta + \gamma = \pi$.

3.1.2 Use the result of Exercise 3.1.1 to show that the angle sum of any quadrilateral (four-sided polygon) is 2π.

3.2 Coordinates

The first great advance in geometry after the time of Euclid was the introduction of *coordinates*: the labelling of points by numbers. We touched on this idea in relation to complex numbers in Chapter 2, but here we study coordinates from scratch, in order to see their interaction with geometry more clearly. Coordinates are a natural consequence of the parallel axiom, so the Greeks were able to think of them, but they could not use them effectively because they did not have algebra. As we have seen, algebra matured only in the sixteenth century, and the first to apply it to geometry were the French mathematicians Pierre de Fermat and René Descartes, around 1630.

The *coordinates* (x, y) of a point P in the plane are simply its distances, horizontal and vertical, from a fixed reference point called their *origin O*. Figure 3.7 shows a few points and their coordinates.

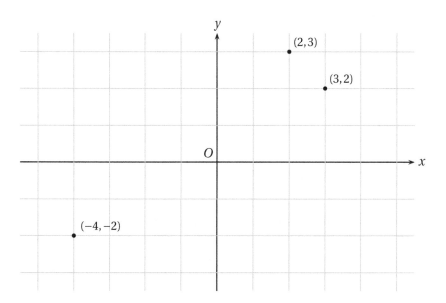

Figure 3.7: Coordinates.

Also shown are the *x-axis*, the horizontal line through *O* which consists of the points $(x, 0)$ for all numbers x; the *y-axis*, which consists of all points of the form $(0, y)$; and the *grid* of horizontal and vertical lines through integer values on the *y-* and *x*-axes, respectively. The grid is like the road grid in certain cities, for example the streets and avenues of Manhattan, and it allows us to name certain points as "street corners," for example the points $(2, 3)$, $(3, 2)$, and $(-4, -2)$ shown.

These "street corners" are *integer points* in the plane, but *any* point *P* has a pair (x, y) of real number coordinates, since *P* has exactly one distance *x* horizontally from *O* and one distance *y* vertically. Since the real numbers represent all points of the line, their pairs represent all points of the plane. The negative numbers are used for points to the left of, or below, *O*.

It is clear from the meaning of *x* and *y* that the point $(2, 3)$ is different from the point $(3, 2)$. (The corner of 2nd Street and 3rd Avenue is different from the corner of 3rd Street and 2nd Avenue.) Thus the order of the *x* and *y* must be respected, and to emphasize this we call (x, y) an *ordered pair*.

Coordinates and the Parallel Axiom

When we say that $P = (a, b)$, or that *P* lies at horizontal distance *a* and vertical distance *b* from *O*, we are implicitly assuming that two different routes lead from *O* to *P*:

- Travel distance *a* horizontally, then distance *b* vertically.

- Travel distance *b* vertically, then distance *a* horizontally.

This amounts to assuming that there is a *rectangle* of any width *a* and any height *b* (Figure 3.8) and, as mentioned in the previous section, rectangles owe their existence to the parallel axiom.

Another concept we owe to the parallel axiom is that of "slope." The gradient or *slope* of a line is measured in the same way that the gradient of a road is measured; by "rise over run." The *rise* from *P* to *Q* is the vertical distance from *P* to *Q*, and the *run* is the horizontal distance from *P* to *Q* (Figure 3.9). For example, a road is said to have a slope of "1 in 10" when the rise is 1 for a run of 10. Mathematically, its slope is the number "1 over 10," that is, $1/10$.

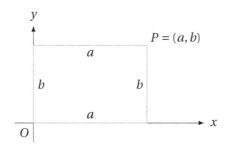

Figure 3.8: Coordinates and rectangle.

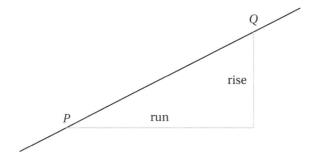

Figure 3.9: Finding the slope of a line.

By calling $\dfrac{\text{rise from } P \text{ to } Q}{\text{run from } P \text{ to } Q}$ the slope of the *line*, not just the slope from P to Q, we assume that this quotient has the same value for any two points P and Q on the line. In other words, *a line has constant slope.* This assumption is equivalent to Euclid's parallel axiom, so it is fair to take it as a fundamental property of lines.

The constant slope of lines enables us to bring algebra into play, because it allows each line to be described by an *equation.* Here is how.

First suppose we have line of slope m, crossing the y-axis at height c as shown in Figure 3.10.

If $P = (x, y)$ is an arbitrary point on the line, we have the rise $y - c$ and run x shown by the gray lines in the figure, and hence

$$\frac{y - c}{x} = \text{slope} = m.$$

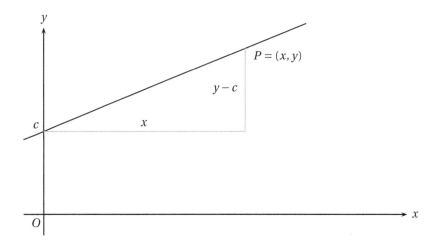

Figure 3.10: The line with equation $y = mx + c$.

Multiplying both sides by x, then adding c to both sides, gives the well-known equation, satisfied by all points on the line,

$$y = mx + c.$$

This argument applies to any line that crosses the y-axis, and hence any such line has an equation of the form $y = mx + c$. If a line does *not* cross the y-axis, it is a vertical line, and hence of the form

$$x = d$$

for some number d.

It is now easy to identify parallel lines from their equations: *parallels are lines with the same slope.* More precisely, a pair of parallels is either of the form

$y = mx + b, \quad y = mx + c$ (both with slope m, but with different b and c)

or of the form

$x = c, \quad x = d$ (both vertical, but with different c and d).

Each of these equation pairs is *inconsistent*—the first implies $b = c$ and the second implies $c = d$—hence neither pair has a solution and the corresponding lines do not meet. Any other pair of equations for lines can be solved, so the equation pairs above are precisely those that represent parallel lines.

We have now described parallel lines geometrically and algebraically, and either way it is clear that they don't meet. So, why would anybody think that they do?

Exercises

3.2.1 On a diagram with x- and y-axes, sketch the lines $y = x$ and $y = x + 1$.

3.2.2 Write down the equation to the line, parallel to those in Exercise 3.2.1, that crosses the y-axis at height 2.

3.2.3 The lines $y = 2x + 1$ and $y = 3x + 2$ have different slopes, so they must have a common point. Find it.

3.2.4 For any two lines $y = mx + c$ and $y = m'x + c'$ with $m \neq m'$, find their common point.

3.3   Parallel Lines and Vision

> What, keep love in *perspective*—that old lie
> Forced on the Imagination by the Eye
> Which, mechanistically controlled, will tell
> How rarely table sides run parallel;
> How distance shortens us; how wheels are found
> Oval in shape far oftener than round;
> How every ceiling corner's out of joint;
> How the broad highway tapers to a point—
> Can all this fool us lovers? Not for long:
> Even the blind will sense that something's wrong.
>
> Robert Graves, In Perspective, *Complete Poems*, vol. 3, p. 133.

... on the boundless Nullabor plain. Before us stretched a single-line track, two parallel bars of shining steel, dead

straight and painfully shiny in the sunshine, and hatched
with endless rungs of concrete ties. Somewhere in the vicin-
ity of the preposterously remote horizon the two gleaming
lines of steel met in a shimmery vanishing point.

Bill Bryson, *In a Sunburned Country*, p. 40.

In the first two sections of this chapter we have discussed parallels from the standpoint of *measurement*: they are lines with the same
slope, or lines a constant distance apart. The passages above remind
us that parallels are quite different from the standpoint of *vision*: if the
plane is spread out horizontally in front of us, parallels appear to meet
on the horizon. Their meeting is "forced on the Imagination by the
Eye."

Yet they still don't actually meet, so what is going on? Well, each line
is infinite and has no endpoint, so the apparent meeting point on the
horizon is not a point on either line. Nevertheless, it is a well-defined
point that rightly "belongs" to each line, so we simply attach it. It is
called the *point at infinity* of the two parallel lines. All lines parallel
to the first two have the same point at infinity, and each point on the
horizon is the point at infinity of some family of parallels. Because of
this, we call the horizon the *line at infinity* of the plane.

Points at infinity complete parallel lines in a natural way, but in any
case the human eye cannot see the difference between a line with an
endpoint and one without. We have tried to indicate the difference
diagrammatically in Figure 3.11 by drawing a "line" at infinity thick
enough to be visible, then erasing it.

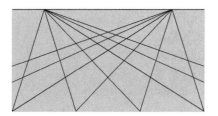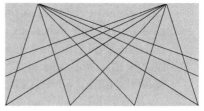

Figure 3.11: The plane with and without points at infinity.

Artists and architects were the first to recognize the importance
of points at infinity, which they call "vanishing points." In fifteenth-

century Italy they discovered how to use them in perspective drawing, spectacularly improving the depiction of three-dimensional structures, as already shown in Figure 3.2.

The mathematical essence of the Italian perspective method is a construction first presented in a treatise on painting by Alberti (1436),[2] for the depiction of a square-tiled floor. His construction assumes that one line of tile edges coincides with the bottom edge of the picture, and chooses any horizontal line as the horizon. Then lines drawn from equally spaced points on the bottom edge to a single point on the horizon depict the parallel lines of tiles perpendicular to the bottom edge (Figure 3.12). Another horizontal line, near the bottom edge, completes the first row of tiles.

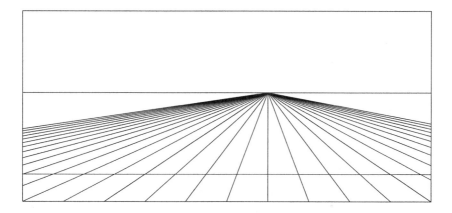

Figure 3.12: Beginning Alberti's construction.

The real problem comes next. How do we find the correct lines to depict the second, third, fourth, ... rows of tiles? The answer is surprisingly simple: draw the *diagonal* of any tile in the bottom row (shown in red in Figure 3.13). The diagonal crosses successive parallels at the corners of tiles in the second, third, fourth, ... rows, so these rows can be constructed by drawing horizontal lines at the successive crossings. Like so:

---

[2]In the first edition of this book I called Alberti's construction the *costruzione legittima* (legitimate construction). I have since learned that this term was not used by Alberti, but by later art historians. So now I simply call it Alberti's construction.

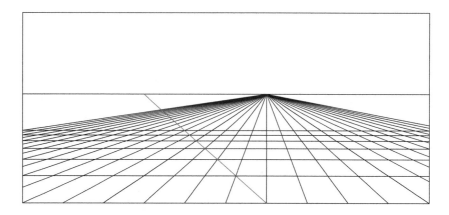

Figure 3.13: Completing Alberti's construction.

Alberti's construction works because certain things remain the same in any view of the plane:

- Straight lines remain straight.

- Intersections remain intersections.

- Parallel lines remain parallel or meet on the horizon.

By choosing to keep "horizontal" lines horizontal, we force the "verticals" to meet on the horizon. And since the intersections of a diagonal with the "horizontals" are also its intersections with the "verticals," the latter intersections give us the positions of the horizontals.

Use by Artists

Italian Renaissance artists seem to have been very content with Alberti's construction. All correct views of tiled floors in their paintings seem to be based on it. With successive rows of tiles lying parallel to the frame, and corners of tiles aligned straight along diagonals, there is a great sense of dignity and order. This no doubt suits the generally sober tone of Renaissance painting, with its themes from religion and classical literature.

A generalization of the construction was given by Jean Pèlerin, a secretary of Louis XI of France, in his book *De artificiali perspectiva*

(1505). This is the first manual on perspective, written in French and Latin, but in fact easy to read from the pictures alone. One of them is shown in Figure 3.14 and the whole book may be perused at the Gallica web site. It was published under the name Viator, the latinized form of the author's name. (Pèlerin means "pilgim" in French, and "viator" is Latin for "traveler.")

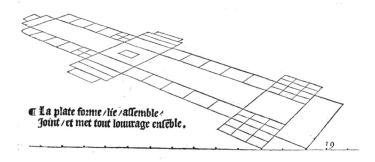

Figure 3.14: Figure from Pèlerin's *De artificiali perspectiva.*

Pèlerin takes the bottom edge of the picture, again marked with a series of equally spaced points, as a diagonal of the tiling. Then the "horizontal" and "vertical" lines are two families of parallels, drawn from the marked points on the diagonal to two points at infinity as in Figure 3.11. The orientation of the two families of parallels is arbitrary, but the view of the tiling itself is not arbitrary, since its left-to-right diagonals are parallel to the horizon.

Pèlerin's scheme offers much greater freedom in the depiction of tiled floors, but artists were slow to take advantage of it. It first became common in Dutch paintings of the seventeenth century, when informal scenes from daily life became popular. Relaxation in the perspective scheme went along with the relaxation in the subject matter. But even then there was not too much relaxation—the two points at infinity were usually placed symmetrically. An example where this is perhaps not true is shown in Figure 3.15.

Figure 3.15: Diagonal tiling in Vermeer's *The Music Lesson*.

Exercises

Figure 3.16 shows some horizontal lines (one of which is the horizon), and some lines meeting on the horizon.

3.3.1 Draw a line on Figure 3.16 (or lay a ruler across it, if you don't want to mark the book) to show that it is *not* a correct perspective view of a path tiled by identical rectangles.

3.3.2 Also find the correct position for the second row of tiles, assuming identical rectangles.

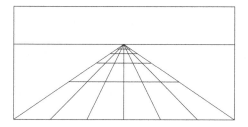

Figure 3.16: Does this path have identical tiles?

3.3.3 Suppose that Figure 3.17 represents a square tile in perspective.

Figure 3.17: A square tile

Explain how to find the correct perspective position of (a) its midpoint, (b) the midpoints of its sides.

3.4 Drawing without Measurement

Drawing the tiled floor in perspective is an important artistic problem and it led to important developments in mathematics. Nevertheless, the cases involving lines parallel to the horizon are mathematically misleading because they are too easily solved with the help of measurement. If the tiled floor has arbitrary orientation then measurement seems useless, and the mind has to focus on the mathematically more fruitful problem of drawing it *without* measurement. How does one draw a view like that shown in Figure 3.18?

This problem is easy to solve, but I do not know whether it was first solved by an artist or a mathematician. The only drawing instrument needed is a straightedge (with no scale marked on it, because measurement is not necessary).

First, a tile is defined by two pairs of parallel lines, that is, two pairs of lines that meet on a line chosen to be the horizon (Figure 3.19).

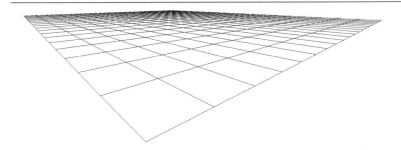

Figure 3.18: Tiled floor with arbitrary orientation.

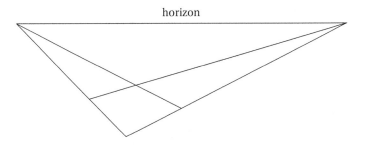

Figure 3.19: Defining a tile.

Then the other tiles are constructed successively, using their diagonals as shown in Figure 3.20. All the diagonals meet at the same point on the horizon.

It is clear that we can continue in this way to construct as many tiles as desired. Only a straightedge is needed, since new lines are constructed by joining points, and new points are constructed by intersecting lines. Mathematically speaking, we assume only the following *incidence axioms* for points and lines:

- There are four points, no three in the same line (a tile to start with).

- There is a unique line through any two points.

- Two lines meet in exactly one point (possibly on the horizon).

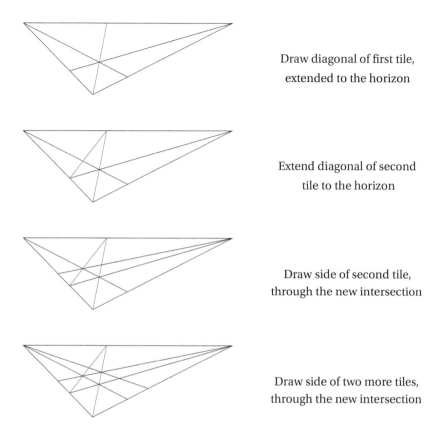

Draw diagonal of first tile,
extended to the horizon

Extend diagonal of second
tile to the horizon

Draw side of second tile,
through the new intersection

Draw side of two more tiles,
through the new intersection

Figure 3.20: Constructing the tiled floor.

Incidence axioms describe the conditions under which objects meet (or are "incident"). The study of them is called incidence geometry, or *projective geometry* because its traditional problems involve projecting a picture from one plane onto another. In particular, drawing a tiled floor in perspective amounts to projecting a grid from the plane of the floor to the plane of the picture.

Projective geometry is in one sense a completion of Euclid's. It adds points at infinity so as to obtain a more homogeneous plane—the *projective plane*—in which parallel lines are no different from other pairs of lines. Parallel lines meet at exactly one point, and the line on which they meet—the horizon—is the same as any other line. In the projective plane we can in fact choose any line h to be the horizon, and the term "parallel lines" means nothing more than "lines that meet on h."

While projective geometry has more homogeneity than Euclidean geometry, it seems to have a smaller stock of concepts. Length and angle are not mentioned, only incidence. However, the three axioms above do not imply all theorems of projective geometry. They do not even explain all the incidence properties of the tiled floor. If we carry out the first few steps in the construction of Figure 3.20 we soon discover that certain miracles occur: three points lie on the same line. A good word for such a miracle is *coincidence*: two points are always incident with a line, but if a third point is incident with the same line as the first two ... that's a *co*incidence. The first such coincidence can be seen in the last step of the figure: three points arising as intersections lie on the same line, shown dashed in Figure 3.21.

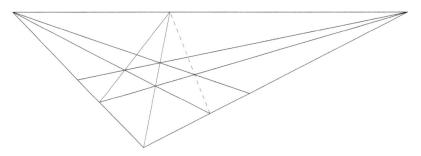

Figure 3.21: A coincidence—three points on a line.

Over the next two sections we shall see that there is a long history of *projective configuration theorems* that explain coincidences. Their proofs lie outside projective geometry, so if we want to explain coincidences within projective geometry we must take certain configuration theorems as *axioms*. But when this is done we experience a new kind of miracle: the concepts of length and angle arise from incidence concepts, so that (as the English mathematician Arthur Cayley put it) "projective geometry is all geometry."

Exercises

Starting with a picture like Figure 3.22, or otherwise:

3.4.1 Draw a perspective view of a floor tiled by identical triangles.

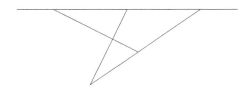

Figure 3.22: Triangle with sides extended to horizon.

3.4.2 Use your picture to construct a perspective view of a floor tiled by identical hexagons.

3.4.3 At which stage in drawing the triangle tiling does the first coincidence occur?

The procedure for drawing a tiled floor with arbitrary orientation leads naturally to a picture of a tiling by triangles, and hence (as Exercise 3.4.2 asks you to show) to a tiling by hexagons. The hexagon tiling is quite common and has been known since ancient times. Yet, strangely, I have not seen any Renaissance paintings of it. The earliest examples I am aware of occur in the book *Floor Decorations of Various Kinds, both in plano and perspective*, published by the English engraver John Carwitham in 1739. The 24 designs by Carwitham may be seen at the web site

    http://blackdoggallery.net/carwitham-designs/.

A particularly beautiful variation of the hexagon tiling is the one shown in Figure 3.23. It is called *rhombille*, presumably because each hexagon is divided into three rhombuses. A 2000-year-old tiling in this pattern exists at Pompeii. The rhombille pattern was also common in Renaissance times, not only on floors, but also on clothing. The first example in Figure 3.23 (the pattern on the gondolier's pants) is from the painting "Miracle of the Holy Cross at the Rialto Bridge" by Vittore Carpaccio, from around 1496. The second is a more recent example of a rhombille floor tiling, in the Zoology Department at the University of Coimbra in Portugal.

3.4.4 Use the perspective view of a triangle tiling to construct a perspective view of rhombille.

Figure 3.23: Examples of rhombille tiling.

The three incidence axioms above are motivated by drawing in the plane, so any structure of "points" and "lines" that satisfies them is called a *projective plane*. However, the "points" and "lines" of a projective plane can be literally *any* things that satisfy the three incidence axioms. In particular, there may be only a finite number of "points" on each "line." The smallest example is the so-called *Fano plane* shown in Figure 3.24.

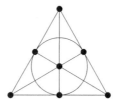

Figure 3.24: The Fano plane.

Here the "points" are the seven black dots and "lines" are the seven curves connecting triples of "points," including the circular "line."

3.4.5 Check that there is a unique "line" through any two "points."

3.4.6 Check that any two "lines" have a unique common "point."

3.4.7 Find four "points," no three of which lie in a "line."

3.5 The Theorems of Pappus and Desargues

The first mathematical book on projective geometry was the *Brouillon project d'une atteinte aux événemens des rencontres du cône avec un plan* (Schematic Sketch of What Happens When a Cone Meets a Plane) of the French engineer Girard Desargues, which appeared in 1639. The book was so obscurely written that it quickly disappeared, and it would still be unknown today but for a lone copy that turned up 200 years later. (What happens when a cone meets a plane, by the way, is that a curve called a *conic section* is produced. This roundabout way of describing the family of curves known as ellipses, parabolas, and hyperbolas is not the least of Desargues' expository failings. Fortunately, he had some followers who ensured that his ideas survived until the time was ripe for them to be appreciated.)

One of Desargues' friends was Etienne Pascal, whose son Blaise later became a giant of French mathematics (and French literature). In 1640, the 16-year-old Blaise became a key contributor to projective geometry, with a theorem about polygons and curves now known as *Pascal's theorem*. Another follower of Desargues was Abraham Bosse, an engraver who expounded some of Desargues' ideas in a manual on perspective for artists in 1648. Bosse's mathematical skills were mediocre, but luckily his book was successful enough to keep Desargues' name alive. Projective geometry finally got a foothold in mathematics with the *Nouvelle méthode en géométrie* (1673) of Phillipe de la Hire, whose father had been a student of Desargues. It seems that de la Hire's book was read by Newton; at any rate, Newton made an important advance by using projection in the geometry of cubic curves. Thus we can say that, by 1700, projective geometry had reached the highest levels of the mathematical world, even if its importance was only dimly recognized.

In those days projective geometry was entangled with Euclid's geometry and the coordinate geometry of Fermat and Descartes. Also, the main applications of projective geometry were thought to be in the theory of curves, as one sees from the title of Desargues' book. Any question about straight lines could be handled by Euclid's geometry, or even more efficiently by coordinates, so the best way to show off projective geometry was by solving problems about curves that were intractable by other means. This accounts for the emphasis on conic sections (by Desargues, Pascal, and de la Hire) and cubic curves (by Newton).

The emphasis on curves is completely understandable, because in Desargues' day only two known theorems about straight lines were interesting from the projective point of view. One of them was discovered by the Greek mathematician Pappus around 300 CE, and the other by Desargues himself.

The Pappus theorem. *If six points A, B, C, D, E, F lie alternately on two straight lines, then the intersections of AB and DE, BC and EF, and CD and FA lie on a straight line.*

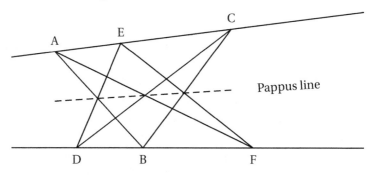

Figure 3.25: The Pappus theorem.

This theorem is interesting because it is purely projective: the first major theorem of this kind ever discovered. Its statement involves only incidence concepts: points lying on lines, lines meeting at points. Despite this, its *proof* requires concepts from Euclid's geometry: lengths of line segments, and also the *product* of lengths. We will see in a moment how multiplication gets involved in the Pappus theorem, but first let us look at Desargues' contribution to the geometry of lines.

The Desargues theorem. *If two triangles are in perspective from a point, then the intersections of their corresponding sides lie on a line.*

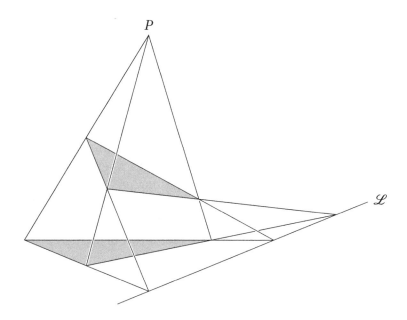

P

$\mathscr{L}$

Figure 3.26: The Desargues theorem.

The gray triangles in Figure 3.26 are in perspective from the point P. That is, the lines through their corresponding vertices pass through P. Under these conditions their corresponding sides meet on a line $\mathscr{L}$.

Like the Pappus theorem, the Desargues theorem is about incidence yet its proof uses length and multiplication. But with an interesting difference: the Desargues theorem also holds if the figure does not lie in a plane, and *this spatial version of the Desargues theorem has a natural projective proof.*

Namely, suppose the triangles in Figure 3.26 are actually in two different planes. Just as two lines in a projective plane meet in a point, *two planes in projective space meet in a line* (possibly a "line at infinity"). Let $\mathscr{L}$ be this line. Then corresponding sides meet on $\mathscr{L}$, if they meet at all, since $\mathscr{L}$ contains all the points common to the two planes. And corresponding sides do meet, because they lie in the same plane— the plane joining them to the point P.

A Proof of the Pappus Theorem

Unfortunately, projective space does not help us with the Pappus theorem. If we want to prove it there is no avoiding concepts from outside projective geometry. However, projective ideas simplify the *statement* of the theorem, and this makes it easier to prove.

Recall from Section 3.4 that we are licensed to call any line in the projective plane the line at infinity. We now invoke this license and let the line at infinity be where AB meets DE and where BC meets EF. In other words, *we assume that AB is parallel to DE and that BC is parallel to EF.* Then we have to show that CD also meets FA at infinity, in other words that *CD and FA are also parallel.*

Figure 3.27 shows the Pappus configuration from this new perspective.

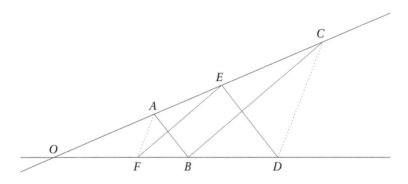

Figure 3.27: The Pappus configuration with Pappus line at infinity.

We are assuming that the colored pairs of lines are parallels, and want to show that the dotted pair are also parallel. To do this we use a consequence of Euclid's parallel axiom: *in triangles of the same shape, corresponding sides have lengths in a constant ratio.*

For example, the triangles OAB and OED have the same shape, because AB is parallel to ED. It follows that

$$\frac{\text{top side}}{\text{bottom side}} = \frac{OA}{OB} = \frac{OE}{OD}. \tag{1}$$

Similarly, triangles OEF and OCB have the same shape because EF is parallel to CB, hence

$$\frac{\text{top side}}{\text{bottom side}} = \frac{OE}{OF} = \frac{OC}{OB}. \tag{2}$$

Multiplying both sides of Equation (1) by $OB \cdot OD$ gives

$$OA \cdot OD = OB \cdot OE, \tag{3}$$

and multiplying both sides of Equation (2) by $OB \cdot OF$ gives

$$OB \cdot OE = OC \cdot OF. \tag{4}$$

From Equations (3) and (4) we have two terms that equal $OB \cdot OE$, namely $OA \cdot OD = OC \cdot OF$. Dividing both sides of the latter equation by $OF \cdot OD$ finally gives

$$\frac{OA}{OF} = \frac{OC}{OD}. \tag{5}$$

This says that triangles OAF and OCD have the same shape, and hence that AF is parallel to CD, as required.   □

(From now on we will occasionally use the □ sign to indicate the end of a proof.) Thus the Pappus theorem has been proved, but not within projective geometry. In effect, we have translated it into Euclid's world, where parallel lines have *slopes*, which are ratios of lengths. If we want the Pappus theorem in the projective world, where parallel lines do not have slopes, we have to take it as an axiom. The same applies to the Desargues theorem. In the next section we will see the first fruits of these new axioms: they explain coincidences.

Exercises

3.5.1 State a parallel version of the Desargues theorem and illustrate it with a diagram.

3.5.2 Prove your parallel version of the Desargues theorem, using similar triangles and proportions.

3.6   The Little Desargues Theorem

The Pappus and Desargues theorems show that certain coincidences—three points lying on the same line—are in fact inevitable. In fact, all such coincidences can be explained as consequences of these two theorems, though this became known only about 100 years ago. The coincidences in the tiled floor actually follow from a special case of the Desargues theorem, as was discovered in the early 1930s by the German mathematician Ruth Moufang.

To illustrate these remarkable discoveries we take the coincidence of Figure 3.21, and show how it follows from the so-called *little Desargues theorem*, which is the special case where the center of perspective *P* lies on the same line $\mathscr{L}$ as the intersections of corresponding sides. Figure 3.28 shows the little Desargues configuration with $\mathscr{L}$ as the horizontal line at the top.

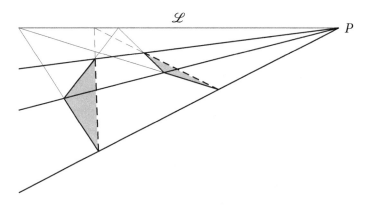

Figure 3.28: The little Desargues theorem.

The theorem states that if two pairs of corresponding sides (solid lines) meet on $\mathscr{L}$ then the third pair (dashed lines) also meet on $\mathscr{L}$. Using our license to view $\mathscr{L}$ as the line at infinity, we can draw the lines meeting on $\mathscr{L}$ as parallels and the lines emanating from *P* also as parallels. This gives the more friendly view of the little Desargues configuration shown in Figure 3.29: $\mathscr{L}$ and *P* have vanished to infinity and lines meeting on $\mathscr{L}$ are *parallel*.

From this viewpoint, the little Desargues theorem simply says: *if*

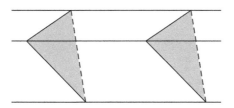

Figure 3.29: Parallel version of the little Desargues theorem.

two triangles have corresponding vertices on parallel lines, and if two pairs of corresponding sides are parallel, then the third pair is also parallel.

Indeed, it would be easy to prove this version of the little Desargues theorem by using slopes, as we did for the Pappus theorem. But this is not our aim. Instead, we wish to see what the little Desargues theorem *implies* about other structures involving parallel lines, such as tiled floors. If it implies certain coincidences, then it implies the same coincidences in the perspective view, where the concept of slope does not apply. In this sense, the little Desargues theorem gives a *projective explanation* of coincidences such as the one in Figure 3.21.

To see the role of the little Desargues theorem more clearly, redraw Figure 3.21 with the parallel lines actually parallel, following the steps in Figure 3.20. You will notice that the new steps are harder to carry out than the old, since parallels have to be drawn. But parallels give a picture, Figure 3.30, in which it is easier to spot instances of the little Desargues theorem. Instead of a line through three points (one of them at infinity) we have a line through two points but parallel to two other lines. The red lines are parallel by construction, but the dashed red line is the diagonal of a constructed tile, and hence parallel to the first solid red lines by coincidence.

The solid black lines are also (two families of) parallels by construction, so there are many opportunities to apply the little Desargues theorem.

First we apply it to the two gray triangles shown in Figure 3.31. Two pairs of corresponding sides (solid black and solid red) are parallel by construction, and the vertices lie on the parallel horizontal lines, hence the dashed black sides are parallel by little Desargues.

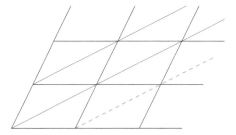

Figure 3.30: Coincidental parallel lines.

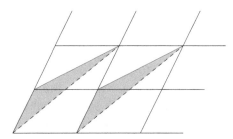

Figure 3.31: First application of little Desargues.

Now we apply it to the two gray triangles shown in Figure 3.32. The solid black sides are parallels by construction and the dashed black sides are parallel by what we have just proved. Again, the vertices lie on the parallel horizontal lines, hence the red sides are parallel by the little Desargues theorem.

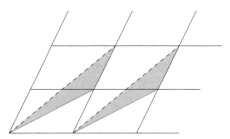

Figure 3.32: Second application of little Desargues.

But the red sides are precisely the lines we wished to prove parallel at the beginning.   □

Despite this projective explanation for coincidental parallels, one might still feel that the "real" explanation should involve lengths. Why assume axioms like the Pappus and Desargues theorems if the same results follow by multiplying and dividing lengths (as in the proof of the Pappus theorem in Section 3.5)?

To some extent this is a question of taste. Most mathematicians are fluent in basic algebra and they find the Pappus and Desargues theorems awkward by comparison. However, the Pappus and Desargues theorems do more than explain projective coincidences—they also explain where basic algebra comes from! This amazing claim is supported by a long series of works by German geometers, notably Christian von Staudt in 1847, David Hilbert in 1899, and Ruth Moufang in 1932. They discovered that *laws of algebra arise from projective coincidences*, and we tell the gist of their story in the next section.

Exercises

As we saw in the exercises to Section 3.4, there are examples of projective planes quite unlike the visual plane that originally motivated the three incidence axioms. We should therefore not expect that the coincidences of perspective drawing occur in all projective planes. Surprisingly, they *do* occur in the Fano plane, so this plane is not as bizarre as it seemed at first.

Perhaps the most contrarian projective plane, which satisfies virtually no nice theorems except the three incidence axioms, is known as the *Moulton plane*. The "points" of the Moulton plane are just the points of the ordinary plane, together with points at infinity. The "lines" are more peculiar. They include the ordinary lines of zero or negative slope but *not* the lines of positive slope. Instead, each other "line" is a bent line whose slope above the x-axis is one half of its slope below the x-axis. Figure 3.33 shows some of the "lines."

3.6.1 Explain why there is a unique "line" through any two "points."

3.6.2 Explain why any two "lines" meet in a unique "point."

3.6.3 Find four "points," no three of which lie on a "line."

3.6.4 By referring to Figure 3.34, explain why the little Desargues theorem fails in the Moulton plane.

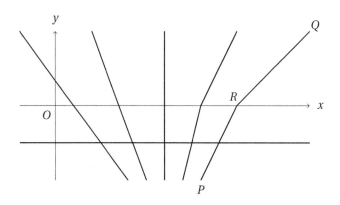

Figure 3.33: "Lines" in the Moulton plane.

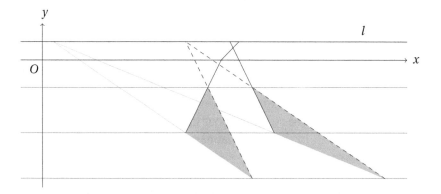

Figure 3.34: Failure of the little Desargues theorem in the Moulton plane.

3.7 What Are the Laws of Algebra?

I was just going to say, when I was interrupted, that one of the many ways of classifying minds is under the heads of arithmetical and algebraical intellects. All economical and practical wisdom is an extension or variation of the following arithmetical formula: $2 + 2 = 4$. Every philosophical proposition has the more general character of the expression $a + b = c$. We are mere operatives, empirics, and egotists, until we learn to think in letters instead of figures.

Oliver Wendell Holmes, *The Autocrat of the Breakfast Table*, p. 1.

Calculating with letters instead of numbers is a big step forward in everyone's education. It is rightly appreciated as a step from the concrete to the abstract, from the particular to the general, from arithmetic to algebra; but it is not always recognized as a step from confusion to clarity. To appreciate the clarity of algebra, ask yourself: what are the rules for calculating with numbers?

First there are some rules specific to the choice of 10 as the base for our numerals: the addition table, the multiplication table, and the rules for "carrying" or "borrowing." These take about a page to write down. Then there are some general rules, such as "in whatever order numbers are added, the result is the same," which might fill another page (it's hard to tell, since these rules are seldom written down).

Compare this with the rules for calculating with letters. Since the letters stand for numbers, they obey the same rules, though rules for a specific base are no longer needed. More importantly, the vague and wordy general rules are replaced by crisp symbolic statements, such as

$$a + b = b + a$$

(which says "in whatever order numbers are added, the result is the same"). This is an astounding advance in simplicity and clarity, at the very low cost of learning to read symbols as well as words.

A complete set of rules for calculating with letters, which we call the *laws of algebra*, may be written in five lines. There are actually nine laws, but they fall naturally into four pairs of corresponding laws for addition and multiplication, and one law linking the two. Here is the complete set:

$$
\begin{array}{lll}
a + b = b + a & ab = ba & \text{(commutative laws)} \\
a + (b + c) = (a + b) + c & a(bc) = (ab)c & \text{(associative laws)} \\
a + 0 = a & a1 = a & \text{(identity laws)} \\
a + (-a) = 0 & \text{for } a \neq 0, \quad aa^{-1} = 1 & \text{(inverse laws)} \\
a(b + c) = ab + ac & & \text{(distributive law)}
\end{array}
$$

In practice, we use some abbreviations involving numbers other than 0 and 1, such as $2a$ for $a + a$, a^2 for aa, a^3 for aaa, and so on. This redeploys the rules specific to base 10 numerals, but they are not an essential part of the system.

A few comments on the meaning of these laws are probably useful:

- The commutative laws say that the *order* in which numbers are added (or multiplied) does not matter.

- The associative laws say that the *grouping* of terms in a sum (or product) of three terms does not matter. As a result, parentheses are unnecessary in a sum of three terms: since $a + (b + c)$ and $(a + b) + c$ are the same, each can be written $a + b + c$. It follows that parentheses are unnecessary in a sum (or product) of any number of terms.

- The identity laws say that 0 is the *additive identity*, the number whose addition has no effect, and that 1 is the *multiplicative identity*. We can say "the" additive identity and "the" multiplicative because the other laws allow us to prove that there is only one of each with the "do nothing" property (see below).

- The inverse laws say that $-a$ is the *additive inverse* of a, the number whose addition to a produces the identity for addition, and that a^{-1} is the *multiplicative inverse* of a when $a \neq 0$ (when $a = 0$, $a^{-1} = 1/a$ does not exist). We usually write $a + (-b)$ as $a - b$ (called "a minus b") and ab^{-1} as a/b or $\frac{a}{b}$ (called "a divided by b"). Again, we can say "the" inverse because the other laws allow us to prove there is only one.

- The distributive law allows a product of sums to be rewritten as a sum of products. In Chapter 2 we saw how the distributive law is the key to explaining why $(-1)(-1) = 1$.

The laws of algebra were condensed to this short list only in the 1830s, by various mathematicians in Great Britain and Germany. A lot of thought went into making the list as short and simple as possible, as one learns by trying to prove some other well-known properties of numbers using only the laws above. They are not obvious! Here are two examples, which we prove by repeatedly using laws of algebra to replace one term by another:

Uniqueness of additive identity. *If e is a number with the identity property of 0, namely $a + e = a$, then in fact $e = 0$.* This follows by adding $-a$ to both sides of the equation $a = a + e$, which gives

$$
\begin{aligned}
0 = a + (-a) = (a + e) + (-a) & \\
= (-a) + (a + e) & \quad \text{by commutative law} \\
= ((-a) + a) + e & \quad \text{by associative law} \\
= (a + (-a)) + e & \quad \text{by commutative law} \\
= 0 + e & \quad \text{by inverse law} \\
= e + 0 & \quad \text{by commutative law} \\
= e & \quad \text{by identity law.}
\end{aligned}
$$

Multiplicative property of zero. *Multiplication by zero causes annihilation; that is, $a0 = 0$.* This follows with the help of uniqueness if additive inverse, because:

$$
\begin{aligned}
a + a0 = a1 + a0 & \quad \text{by identity law} \\
= a(1 + 0) & \quad \text{by distributive law} \\
= a1 & \quad \text{by identity law} \\
= a & \quad \text{by identity law.}
\end{aligned}
$$

But $a + a0 = a$ says that $a0$ is an additive identity, and hence it equals 0 by the proposition just proved. Notice that we used all five types of law to prove this well-known property of zero!

It was realized from the sixteenth-century beginnings of algebra that calculation with letters explains general properties of calculation with numbers. A pattern *observed* in making many calculations with numbers can be *proved* by a single calculation with letters.

For example, the equations

$$1 \times 3 = 3 = 2^2 - 1$$
$$2 \times 4 = 8 = 3^2 - 1$$
$$3 \times 5 = 15 = 4^2 - 1$$
$$4 \times 6 = 24 = 5^2 - 1$$
$$\vdots$$

are all instances of the pattern $(a-1)(a+1) = a^2 - 1$. You probably know how to prove that $(a-1)(a+1) = a^2 - 1$, and will object to seeing the calculation in tedious detail. However, I want to show exactly how the laws of algebra come into play, so here it is. (The penultimate step, replacing $a0$ by 0, is justified by the series of steps proving the multiplicative property of 0 above.)

$(a-1)(a+1) = (a-1)a + (a-1)1$	by the distributive law
$= (a-1)a + (a-1)$	by the identity law
$= (a-1)a + a - 1$	by the associative law
$= a(a-1) + a - 1$	by the commutative law
$= a^2 + a(-1) + a - 1$	by the distributive law
$= a^2 + a(-1) + a1 - 1$	by the identity law
$= a^2 + a((-1) + 1) - 1$	by the distributive law
$= a^2 + a(1 + (-1)) - 1$	by the commutative law
$= a^2 + a0 - 1$	by the inverse law
$= a^2 + 0 - 1$	by multiplicative property of zero
$= a^2 - 1$	by the identity law

Calculation with numbers is the obvious model for calculation with letters, but a geometric model is also conceivable, since numbers can be interpreted as lengths. Indeed, the coordinate geometry of Fermat and Descartes was based on algebra. They found that the curves studied by the Greeks can be represented by equations, and that algebra unlocks their secrets more easily and systematically than classical geometry. But to apply algebra in the first place, Fermat and Descartes assumed classical geometry. In particular, they used Euclid's parallel axiom and the concept of length to derive the equation of a straight line, as we did in the early sections of this chapter.

A new approach to both algebra and geometry was developed in the nineteenth and twentieth centuries. In 1847, Christian von Staudt found a way to model algebra in geometry *without* using the concept of length. He defined addition and multiplication with the help of parallel lines, but in projective geometry where "parallel" simply means "meeting on a line designated as the line at infinity." This avoids using the concept of length, but raises the new problem of showing that laws of algebra hold for the addition and multiplication so defined.

The sums $a+b$ and $b+a$, for example, represent points obtained by two different projective constructions: one "adding point a to point b," and the other "adding point b to point a." We want the two resulting points to coincide, which is a projective *coincidence*. Indeed, all the laws of algebra correspond to projective coincidences, and von Staudt showed that *all the required coincidences follow from the theorems of Pappus and Desargues.*

Then in 1899 David Hilbert showed that all laws of algebra except the commutative law for multiplication follow from the Desargues theorem. And in 1932 Ruth Moufang showed that all except the commutative and associative laws follow from the little Desargues theorem. Thus the Pappus, Desargues, and little Desargues theorems are mysteriously aligned with the laws of multiplication!

We investigate this alignment between geometry and algebra further in the next section.

Exercises

3.7.1 Prove the uniqueness of multiplicative identity—if $a \cdot e = a$ then $e = 1$—along the lines of the proof above for uniqueness of additive identity. (Try changing all $+$ signs to multiplication signs. What else do you have to change?)

3.7.2 Prove the uniqueness of additive inverse—if $a + e = 0$ then $e = -a$—by adding $-a$ to both sides of the equation $0 = a + e$ and deducing the equation $-a = e$.

The laws of algebra hold for addition and multiplication of numbers, as we know, but they also hold for some less familiar objects that can be added and multiplied. An example is a system of two objects, which we will call **0** and **1**, added and multiplied according to the following rules:

$$0+0=0, \quad 0+1=1+0=1, \quad 1+1=0,$$
$$0 \cdot 0 = 0, \quad 0 \cdot 1 = 1 \cdot 0 = 0, \quad 1 \cdot 1 = 1. \qquad (*)$$

3.7.3 Check that **0** and **1** satisfy all the laws of algebra when added and multiplied by these rules.

We can interpret **0** and **1** as the even and odd classes of natural numbers, in which case the above rules for addition and multiplication express familiar facts such as "even + even = even."

3.7.4 Writing even numbers as $2p$ or $2q$, and odd numbers as $2p+1$ or $2q+1$, check that odd · odd = odd by calculating $(2p+1)(2q+1)$ and showing that the result equals 2·something +1. Also check the other rules (*).

In retrospect, it is obvious that "even" and "odd" satisfy most of the laws of algebra, because natural numbers satisfy all the laws except the inverse law for multiplication. This inverse law holds in the system of "even" and "odd" because "odd · odd = odd," which shows that "odd" is its own inverse.

3.7.5 Check that **0**, **1**, and **2** satisfy the laws of algebra when

0 means "multiple of 3,"

1 means "1+multiple of 3,"

2 means "2+multiple of 3."

3.7.6 Show in particular that, under this interpretation, $2 \cdot 2 = 1$ so $2^{-1} = 2$.

3.8　Projective Addition and Multiplication

It is easy to see how to add and multiply points if we first pretend that they represent distances from the origin, and make use of parallel lines to shift and magnify distances. After we have worked out constructions for addition and multiplication, we can drop the pretense, and we still have *definitions* for addition and multiplication of points on a line in a projective plane.

To add point a to point b we use the construction in Figure 3.35.

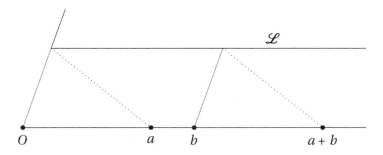

Figure 3.35: Adding a to b.

The points O, a, and b are on a line called the x-axis, and any other line through O is chosen as the y-axis. We also need a parallel $\mathscr{L}$ to the x-axis, though only as "scaffolding," since $a + b$ does not depend on which $\mathscr{L}$ we choose.

To find $a + b$ we travel from b to $\mathscr{L}$ on a parallel to the y-axis (red), then back to the x-axis on a parallel to the line from a to $\mathscr{L}$ and the y-axis (dotted).

In Euclid's geometry this creates a pair of congruent triangles, each with base length a, so $a + b$ is an appropriate name for the point arrived at. In projective geometry, however, we cannot appeal to the concept of length, so we have to *prove* that $a + b$ has the properties expected of the sum of a and b.

For example, does $a + b = b + a$? To construct $b + a$ we reverse the roles of a and b in the construction above, obtaining Figure 3.36.

This is a different construction. But the two constructions lead to the same point, by a projective coincidence that is easily seen when we superimpose the two pictures (Figure 3.37): the theorem of Pappus!

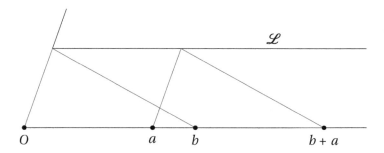

Figure 3.36: Adding b to a.

The point $a + b$ lies at the end of the dotted line (parallel to the other dotted line), the point $b + a$ lies at the end of the blue line (parallel to the other blue line), and Pappus says these are the same point.

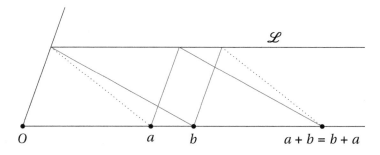

Figure 3.37: Why $a + b = b + a$.

To multiply a by b we use the same set-up, with O, a, and b on the x-axis and another line through O called the y-axis. Now we also need a point called 1 on the x-axis, different from O, and we construct corresponding points called 1, a, and b on the y-axis by connecting them to 1, a, and b on the x-axis by a family of parallels (shown in red). Now consider the triangles cut off from the region between the x- and y-axes by the dotted lines shown in Figure 3.38. These lines are

- the line from 1 on the y-axis to a on the x-axis, and

- the parallel to it from b on the y-axis.

In Euclid's geometry, parallel lines cut off similar triangles, and the sides of the second are a magnification of the corresponding sides of the first by b. Thus it is reasonable to define the other end of the dotted line from b to be ab.

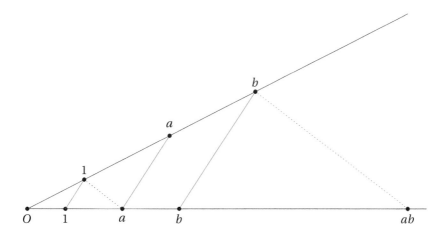

Figure 3.38: Multiplying a by b.

When we reverse the roles of b and a the construction is different, so the result ba is conceivably different from ab. Drawing the construction lines now blue instead of dotted, we find the point ba at the other end of a blue line from a on the y-axis, parallel to a blue line from 1 on the y-axis to b on the x-axis (Figure 3.39).

It turns out that $ab = ba$ all right, by exactly the same projective coincidence that shows $a + b = b + a$: the Pappus theorem. This can be seen when we superimpose the construction of ba on the construction of ab, which again brings the Pappus configuration to light (Figure 3.40). By Pappus, the end ab of the dotted line is the same as the end ba of the blue line.

The Pappus theorem seems tailor-made to show $a + b = b + a$ and $ab = ba$, but other laws of algebra are more easily derived from the Desargues theorem. For example, Hilbert (1899) showed that the associative law for multiplication follows from Desargues. Rather surprisingly, it was not noticed until 1904 that Desargues follows from Pappus, and hence that all the laws of algebra follow from Pappus alone.

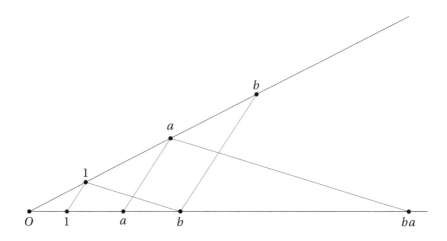

Figure 3.39: Multiplying b by a.

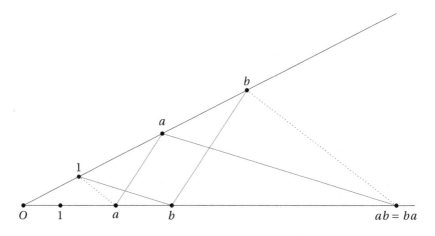

Figure 3.40: Why $ab = ba$.

What Are the Axioms of Geometry?

Euclid's geometry was based on a small number of axioms, the most important of which was the parallel axiom. However, Euclid also made many unconscious assumptions, and by the time these were recognized (around 1900, by Hilbert and others) the number of Euclid-style axioms had grown to around 20. This is more than the number of laws of algebra, even when we add axioms to describe the behavior of points (the so-called *vector space* axioms) and length (the so-called *inner product* axioms). Also, of course, the laws of algebra apply in other parts of mathematics. Thus by the mid-twentieth century a view developed that algebra was the proper foundation for geometry.

But the results sketched above show that most of Euclid's geometry *and* the laws of algebra follow from just four axioms of projective geometry:

1. There are four points, no three of which are in a line.

2. Any two points lie in a unique line.

3. Any two lines meet in a unique point.

4. The Pappus theorem.

This discovery from 100 years ago seems capable of turning mathematics upside down, though it has not yet been fully absorbed by the mathematical community. Not only does it defy the trend of turning geometry into algebra, it suggests that both geometry and algebra have a simpler foundation than previously thought.

Those searching for signs of intelligent life elsewhere in the universe have often recommended looking for signs of mathematical ideas, such as prime numbers or the Pythagorean theorem, in the stream of electromagnetic noise reaching the earth. But the results above make it less clear that we will be able to recognize alien geometry or algebra when it is transmitted to us. At the very least, we should look for signs of the Pappus theorem as well!

Exercises

3.8.1 Prove $a + O = a$ directly from the definition of projective addition. (It is a very short proof!)

3.8.2 Show also that there is a point $-a$ such that $a + (-a) = O$.

3.8.3 Prove $a1 = a$ from the definition of projective multiplication.

Chapter 4

The Infinitesimal

Preview

Thanks to the real numbers, we can measure the length of any line, even irrational lines such as the diagonal of the unit square. A different problem arises when we want to measure the length of curves. In the *Geometry* of René Descartes [14, p. 91] of 1637 we find the famous prediction:

> ... the ratios between straight and curved lines are not known, and I believe cannot be known by human minds.

In fact, even the straight figures of two and three dimensions (polygons and polyhedra) are hard to measure. With some ingenuity, it is possible to measure the area of any polygon by cutting it into finitely many pieces and reassembling them to form a square. But to find the volumes of polyhedra it sometimes seems necessary to cut them into *infinitely many* pieces, and hence into pieces that are arbitrarily small.

Now "arbitrarily small pieces" simply means that, for any assigned size, there is a piece smaller than that size. It does *not* mean that there is one piece smaller than any size. Such a piece is called *infinitesimal*, and it is surely impossible. *Yet, despite their impossibility, infinitesimals are easy to work with and they usually give correct results.*

In the present chapter we will review the basic theory of length, area, and volume, and see how *infinitesimals insinuate themselves into the study of curved figures.* In finding tangents, areas, and the length of curves such as the circle, *it seems that the shortest route to the truth passes through the impossible.*

4.1 Length and Area

One of the fundamentals of mathematics is the relationship between length and area. We have already seen how the Pythagorean theorem relates the two. A more elementary, but crucial, relationship is seen in the formula:

$$\text{area of a rectangle} = \text{base} \times \text{height},$$

where the "base" is the length of one side (conventionally shown as horizontal in pictures) and the "height" is the length of the perpendicular side.

This can be taken as the *definition* of area for rectangles, but it is motivated by the case where the base is m units and the height is n units, for some integers m and n. In this case, the rectangle clearly consists of mn unit squares, as Figure 4.1 shows in the case $m = 5$ and $n = 3$:

Figure 4.1: Area of a rectangle.

Since we think of lengths as numbers, and the product of numbers is a number, we think of areas as numbers as well. The Greeks thought length was a more general concept than number, because length can be irrational, so they viewed the product of lengths literally as a rectangle. This raises the question of when two rectangles are "equal" (that is, equal in area in our language), which Euclid answered by *cutting and pasting.* He called polygons $\mathscr{P}$ and $\mathscr{Q}$ *equal* if $\mathscr{P}$ can be cut by straight lines into pieces that can be reassembled to form $\mathscr{Q}$.

Even today, when we are happy to multiply irrationals, the area of a general polygon is best found by cutting and pasting it to form a rectangle. We need to cut and paste simply to find *which* numbers to multiply. For example, any parallelogram can be converted to a rectangle by cutting a triangle off one end and pasting it onto the other as in Figure 4.2.

Figure 4.2: Cutting and pasting a parallelogram.

It follows that

$$\text{area of parallelogram} = \text{base} \times \text{height},$$

where the "base" is the length of one side and the "height" is the distance between the base and its parallel side.

Next we find that

$$\text{area of triangle} = \frac{1}{2}\,\text{base} \times \text{height}.$$

because any triangle is half of a parallelogram with the same base and height, as Figure 4.3 shows.

$$= \quad \frac{1}{2}$$

Figure 4.3: Triangle as half a parallelogram.

Finally, it is possible to complete the story of area for polygons by showing that

- any polygon may be cut into triangles, and

- the area of a polygon (that is, the area of any rectangle obtained from it by cutting and pasting) is the sum of the areas of these triangles.

These facts are *not* entirely obvious, though it would be shocking if they were false. The second fact is so far from obvious that it was first proved by the great David Hilbert as late as 1898. We leave you to think about both facts, but we do not require their proofs, since we do not need to know the area of any other polygons.

Exercises

4.1.1 Consider the $(3, 4, 5)$ triangle.

- If the side of length 3 is the "base," what is the height?
- If the side of length 4 is the "base," what is the height?
- If the side of length 5 is the "base," what is the height?

4.1.2 Explain how to cut a triangle into pieces that reassemble to form a rectangle with the same base but half the height of the triangle. (You may take whichever side is most convenient for the "base.")

4.1.3 If we take the side of length 13 in the $(5, 12, 13)$ triangle as the "base," what is the height?

4.2 Volume

Cutting and pasting also works with volume, but not so completely. The analogue of a rectangle is a *box*, and consideration of unit cubes (as in Figure 4.4) prompts us to make the definition

$$\text{volume of box} = \text{length} \times \text{width} \times \text{height}$$
$$= \text{base} \times \text{height},$$

where "base" now denotes the base area, length × width.

The analogue of a parallelogram is called a *parallelepiped*, a solid whose opposite faces are parallel. (To pronounce its name correctly, break it down as "parallel-epi-ped." These are the Greek words that express its meaning—"parallel-upon-foot.") It is a nice visualization

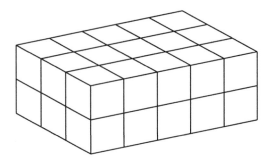

Figure 4.4: Volume of a box.

exercise to see that any parallelepiped can be cut and pasted into a box
with the same base area and height. (Have a good look at Figure 4.5.)

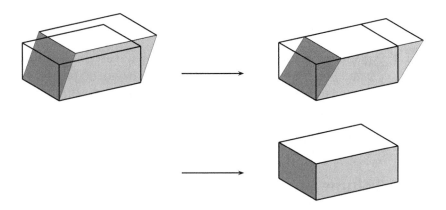

Figure 4.5: Parallelepiped and box.

Thus we conclude that

volume of parallelepiped = base × height.

If we cut a parallelepiped in half through parallel diagonals of the
top and bottom faces we get a figure called a *triangular prism.* Its volume is half that of the parallelepiped, but its base is also halved, hence

volume of triangular prism = base × height.

By pasting triangular prisms together, we can make a *generalized prism*, with an arbitrary polygon as its horizontal cross section, and its volume is also given by the formula base × height.

However, cutting and pasting fails with even the simplest solid figure not of constant cross section—the tetrahedron, or triangular pyramid. All methods for finding the volume of a general tetrahedron involve infinite processes, and we shall study one of them in the next section.

Exercises

4.2.1 Find the volume of a prism of height 7 whose base is a $(3, 4, 5)$ triangle.

4.2.2 What is the side length of a cube with volume equal to 8?

4.2.3 What is the side length of a cube with volume equal to 2?

4.3 Volume of a Tetrahedron

Euclid's derivation of the volume of a tetrahedron is in the *Elements*, Book XII, Proposition 4, and it is based on the diagram below.

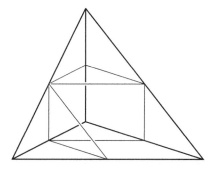

Figure 4.6: Euclid's subdivision of the tetrahedron.

Figure 4.6 shows Euclid's starting point, a subdivision of the tetrahedron by certain lines connecting midpoints of the edges. These lines create two triangular prisms inside the tetrahedron:

- An "upright" prism whose height is half the height of the tetrahedron and whose base is one quarter of the base of the tetrahedron. (In fact the prism's base is one of four congruent triangles
  into which the tetrahedron base is divided by lines joining the
  midpoints of its sides.) We therefore have

$$\text{volume of upright prism} = \frac{1}{8}\,\text{base} \times \text{height},$$

  where "base" and "height" denote the base area and height of the
  tetrahedron.

- A "reclining" prism, lying on a parallelogram that is half the base
  of the tetrahedron (made from two of the four congruent triangles). Viewing this parallelogram as the base of a parallelepiped
  with half the height of the tetrahedron, we see that the reclining
  prism is half this parallelepiped and hence

$$\text{volume of reclining prism} = \frac{1}{8}\,\text{base} \times \text{height}.$$

Thus the volume of the two prisms in the tetrahedron is $\frac{1}{4}$ base × height,
where "base" and "height" (we reiterate) are those of the tetrahedron.

If the two prisms are removed, two half-size tetrahedra remain, and
these may be similarly subdivided (Figure 4.7).

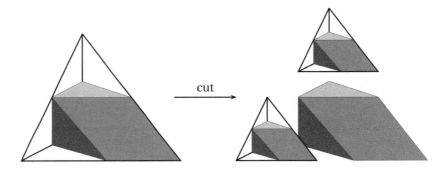

Figure 4.7: Euclid's dissection of the tetrahedron.

The prisms in the half-size tetrahedra are themselves half-size, and
hence 1/8 the volume of the original prisms. Since there are two half-
size tetrahedra, the prisms in them fill 1/4 of the volume of the original

prisms, and hence

$$\text{volume of half-size prisms} = \left(\frac{1}{4}\right)^2 \text{base} \times \text{height},$$

where "base" and "height" again are those of the original, full-size, tetrahedron.

Removing the prisms from the two half-size tetrahedra, we are left with four quarter-size tetrahedra, which are similarly subdivided. This gives

$$\text{volume of quarter-size prisms} = \left(\frac{1}{4}\right)^3 \text{base} \times \text{height}.$$

By continuing the subdivision indefinitely, we fill the entire interior of the tetrahedron with prisms, because each point inside the tetrahedron falls inside some prism. This shows that the volume of the tetrahedron equals the sum of the volumes of the prisms, that is

$$\text{volume of tetrahedron} = \left[\left(\frac{1}{4}\right) + \left(\frac{1}{4}\right)^2 + \left(\frac{1}{4}\right)^3 + \cdots\right] \text{base} \times \text{height}.$$

It remains to evaluate the infinite sum

$$S = \left(\frac{1}{4}\right) + \left(\frac{1}{4}\right)^2 + \left(\frac{1}{4}\right)^3 + \cdots.$$

This can be done by a simple trick. Multiplying both sides by 4, we get

$$4S = 1 + \left(\frac{1}{4}\right) + \left(\frac{1}{4}\right)^2 + \left(\frac{1}{4}\right)^3 + \cdots,$$

and subtracting the first equation from the second gives

$$3S = 1, \quad \text{and therefore} \quad S = \frac{1}{3}.$$

Thus we finally have:

$$\text{volume of tetrahedron} = \frac{1}{3} \text{base} \times \text{height}.$$

Geometric Series

The infinite sum

$$\left(\frac{1}{4}\right) + \left(\frac{1}{4}\right)^2 + \left(\frac{1}{4}\right)^2 + \cdots$$

is an example of the *geometric series*

$$a + ar + ar^2 + ar^3 + \cdots,$$

which has a meaningful sum for any r of absolute value less than 1. The sum can be found by a trick like that used above (and also like the trick used to evaluate periodic decimals in Section 1.6). We put

$$S = a + ar + ar^2 + ar^3 + \cdots$$

and multiply both sides by r to obtain

$$rS = ar + ar^2 + ar^3 + ar^4 + \cdots.$$

Subtracting the second equation from the first gives

$$(1-r)S = a, \quad \text{and therefore} \quad S = \frac{a}{1-r}.$$

This result can be obtained in a more cautious way (which reveals why r must have absolute value less than 1) by using the trick on the *finite* sum

$$S_n = a + ar + ar^2 + ar^3 + \cdots + ar^n.$$

This leads to the result

$$S_n = \frac{a - ar^{n+1}}{1-r}.$$

If we now let n increase indefinitely, the term ar^{n+1} will decrease to zero, *provided r has absolute value less than 1.*

Exercises

4.3.1 Find the sum $\frac{1}{2} + \left(\frac{1}{2}\right)^2 + \left(\frac{1}{2}\right)^3 + \left(\frac{1}{2}\right)^4 + \cdots$

4.3.2 Find the sum $2 + 2^2 + 2^3 + 2^4 + \cdots + 2^n$.

4.3.3 Find the sum $a + ar + ar^2 + \cdots + ar^n$ when $r = 1$.

4.3.4 Show that $\frac{1}{1+x^2} = 1 - x^2 + x^4 - x^6 + \cdots$ when $|x| < 1$. Is this true when $x = 1$?

4.4 The Circle

By cutting and pasting polygons to form rectangles, as in Section 4.1, we can construct a square of the same area as any given polygon. The great challenge of ancient mathematics was to *square the circle*, that is, to construct a square of the same area as the disk bounded by the unit circle. Since the circle is curved it is impossible to actually build a disk from polygons, although its area seems to be a meaningful concept.

Today, we denote the area of the unit disk by π, a number which every school child knows is approximately 22/7. However, π is definitely *not* equal to 22/7. By approximating the circle by polygons of 96 sides, inside and outside, Archimedes was able to show that

$$3\frac{10}{71} < \pi < 3\frac{1}{7},$$

so 22/7 is merely a good approximation to π. It is accurate to two decimal places. The Chinese were also interested in the value of π and Zǔ Chōngzhī (429–500 CE) discovered the remarkable approximation 355/113, which is accurate to six digits. It later became a kind of sporting contest to produce more and more decimal places of π, with 35 places being produced by van Ceulen in 1596, 100 places by Machin in 1706, and 527 by Shanks in 1874 (Shanks actually computed 707 places, but made an error in the 528th place). Shanks was the record holder until the era of electronic computers, and the record now stands at billions of places (perhaps trillions by the time you read this). The computation of π is a great human interest story, but also a great story of human folly, because *the decimal digits of π have so far produced no insight whatsoever.*

It is not of great interest to know, say, the first 50 digits of π,

3.14159265358979323846264338327950288419716939937511 $\cdots$

because these digits give us no idea what the 51st digit is. As with $\sqrt{2}$ there is no apparent pattern, and we do not even know whether any particular digit, say 7, occurs infinitely often. Essentially our only knowledge of the decimal expansion of π is the negative knowledge that it is not periodic, because π is known to be irrational and numbers with periodic decimals are rational, as we saw in Section 1.6.

We have to face the fact that any numerical description of π involves an infinite process, but we can still hope that the process has a clear and simple pattern. This hope is realized by the amazing formula

$$\frac{\pi}{4} = 1 - \frac{1}{3} + \frac{1}{5} - \frac{1}{7} + \frac{1}{9} - \cdots,$$

which is about as simple as any infinite formula can be. It was discovered in India around 1500 CE and rediscovered in Europe in 1670. To explain it, we need some *infinitesimal geometry*, a paradoxical concept that can be introduced by going back to another highlight of Greek mathematics—the relation between length and area of the circle.

As we have seen, the Greeks did not know π exactly, but they did know that the *same value π* is involved in both the length and area of the circle (and in fact also in the surface area and volume of the sphere). You probably remember from school that, for a circle of radius R,

$$\text{length of circumference} = 2\pi R,$$

$$\text{area of disk} = \pi R^2.$$

Around 400 BCE, the Greeks discovered that *if π is defined by* circumference $= 2\pi R$, then *it follows that* area $= \pi R^2$. This relationship between length and area can be seen by imagining the disk cut into a large number of thin sectors and standing the sectors in a line as shown in Figure 4.8. In this illustration we have lined up 20 sectors of angle 18°, but for better results you should imagine 100 sectors, or 1,000, or 10,000, … .

The idea is: if the sectors are very thin, they are virtually triangles whose bases add up to the circumference $2\pi R$ of the circle and whose heights are R. Since the area of each triangle is $\frac{1}{2}$ base × height, it follows that

$$\text{area of disk} = \text{total area of sectors}$$

$$= \frac{1}{2}\text{total base} \times \text{height}$$

$$= \frac{1}{2}2\pi R \times R$$

$$= \pi R^2.$$

But an actual sector, no matter how small, is not precisely a triangle. It bulges slightly at the bottom, so the length of the line of sectors

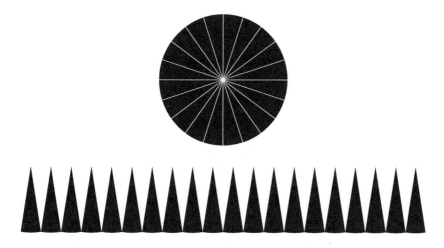

Figure 4.8: Aligning sectors of a disk.

will always fall a little short of the circumference of the circle. (Indeed, the length of the line here is the length of a 20-gon inscribed in the circle.) Likewise, the height of each triangle will always be a little less than R, the length of its sloping side.

Somehow, we see the correct result through *wanting sectors to be something they are not, namely triangles.*

Around 350 BCE Eudoxus found a way to justify conclusions found by this devious kind of argument: the so-called *method of exhaustion.* This method has an echo in the saying of Sherlock Holmes (see Sir Arthur Conan Doyle, *The Sign of Four*, Chapter 6.)

> When you have eliminated the impossible, whatever remains, however improbable, must be the truth.

In the case of the disk it can be shown that the total area of the sectors is never greater than πR^2 and also, by making the sectors sufficiently thin, that their total area exceeds any number less than πR^2. Thus we exhaust all possibilities except the truth, that area $= \pi R^2$.

However, it is tedious to estimate these areas by actually comparing sectors with triangles. In the seventeenth century, mathematicians noticed that they could skip the argument by exhaustion, and easily compute areas and lengths of many curved figures by indulging geometric fantasies such as sectors that behave like triangles.

For example, to show that the area of the disk is πR^2 one imagines the disk divided into *infinitesimal* sectors. These are sectors so thin that *there is no error* in assuming they are triangles of height R whose bases add up to $2\pi R$. This is an audacious and fantastic assumption, but seventeenth-century mathematicians were always able to retort: if you don't believe it, check the result by the method of exhaustion.

Within a few decades of its discovery, *the fantasy of infinitesimals had completely overpowered the honest method of exhaustion* by joining forces with algebra to form an *infinitesimal calculus*—a symbolism for solving problems about curves by routine calculations, like those already known for the geometry of straight lines. The calculus, as we know it today, is perhaps the most powerful mathematical tool ever invented, yet it originated in the dream world of infinitesimals. We look at these strange origins in the next section.

Exercises

4.4.1 Assuming that $\pi = 3.14159625\cdots$, use a calculator to check the result of Archimedes that $3\frac{10}{71} < \pi < 3\frac{1}{7}$.

4.4.2 Check that Zǔ Chōngzhi's number $\frac{355}{113}$ agrees with π to six digits.

4.4.3 Explain why

$$4\left(1 - \frac{1}{3} + \frac{1}{5} - \frac{1}{7} + \frac{1}{9} - \frac{1}{11} + \frac{1}{13} - \frac{1}{15} + \cdots\right)$$
$$= 8\left(\frac{1}{1 \times 3} + \frac{1}{5 \times 7} + \frac{1}{9 \times 11} + \frac{1}{13 \times 15} + \cdots\right)$$
$$= 8\left(\frac{1}{2^2 - 1} + \frac{1}{6^2 - 1} + \frac{1}{10^2 - 1} + \frac{1}{14^2 - 1} + \cdots\right).$$

4.5   The Parabola

Geometry becomes algebraic when points are given by *coordinates* and curves by *equations*. As explained in Section 3.2, each point in the plane is described by a pair (x, y) of real numbers x and y, which represent the horizontal and vertical distances (respectively) from a point O called the *origin*. (Figure 4.9.)

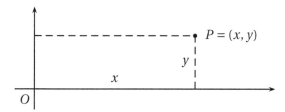

Figure 4.9: A point and its coordinates.

The points on a curve C satisfy some equation called the *equation of C*. For example, the equation

$$y = x$$

represents the line through O with 45° slope, and the equation

$$y = x^2$$

represents a curve through O called a *parabola* (Figure 4.10).

The parabola is one of a classical family of curves known as the *conic sections*, so called because they result from cutting a cone by a plane. They were studied around 200 BCE by the Greek mathematician Apollonius, who found hundreds of their properties by ingenious geometrical reasoning. When Fermat and Descartes introduced coordinates into geometry, around 1630, one of their first steps was to revisit the conic sections. They found that conic sections are precisely the curves given by *quadratic* equations in x and y, so conic sections are algebraically the simplest curves (except for straight lines, which are the curves given by *linear* equations $ax + by = c$ for constants a, b, and c).

Thus coordinates translate the geometry of conic sections into the algebra of quadratic equations. This makes it easy, almost mechanical, to prove most of the properties discovered with great effort by Apollonius. The secrets of more complicated curves, given by cubic equations or worse, are also unlocked, to an extent limited only by the algebraic difficulties of higher degree equations. Some problems can be solved for *all* curves given by algebraic equations. One of these is the *problem of tangents*.

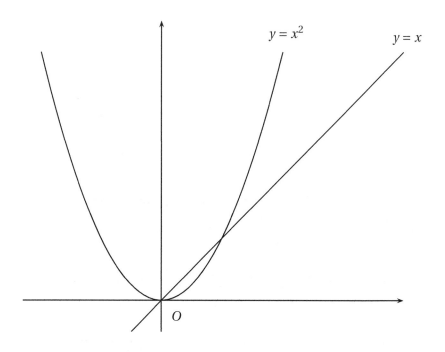

Figure 4.10: Line and parabola.

The *tangent* at a point P on a curve C may be defined as the line through P with the same direction as C at P. In more physical terms, it is the line a moving particle would follow if it left C at point P (that is, if it "went off on a tangent"). Finding the tangent at P therefore amounts to finding the *slope* of the curve at P. We first solve this problem for the parabola $y = x^2$ and then indicate how the method extends to many other curves.

To find the tangent at the point $P = (x, y)$ on the parabola $y = x^2$ we consider the line joining P to a point Q on the parabola, infinitesimally close to P. The coordinates of Q are written $(x + dx, y + dy)$, where dx stands for the infinitesimal "difference in x value" and dy stands for the infinitesimal "difference in y value." The d notation for infinitesimals was introduced by Leibniz around 1680. Figure 4.11 shows how dx and dy relate to the curve.

Since the equation of the parabola is $y = x^2$, its height y at P is x^2

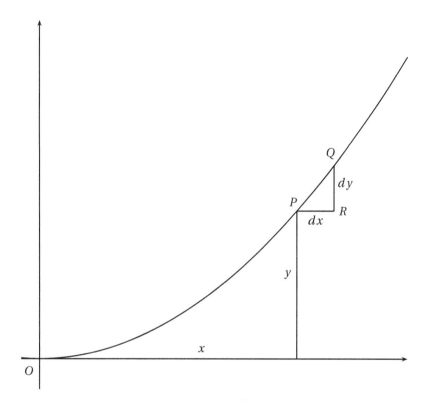

Figure 4.11: Infinitesimally close points on the parabola.

and its height $y + dy$ at Q is $(x + dx)^2$. Therefore,

$$dy = (x + dx)^2 - x^2 = x^2 + 2x\,dx + (dx)^2 - x^2 = 2x\,dx + (dx)^2.$$

Now the slope of the curve at P differs only infinitesimally from the slope of the infinitesimal line PQ, which is dy/dx. From the expression for dy just computed we find that

$$\text{slope of } PQ = \frac{dy}{dx} = 2x + dx.$$

The slope from (x, y) to the infinitesimally close point $(x + dx, y + dy)$ therefore *differs infinitesimally* from $2x$. The true value of the slope *at* (x, y) must therefore be exactly $2x$, that is

$$\text{slope at } P = 2x.$$

Exercises

It is possible to find tangents to certain curves without using infinitesimals. The alternative is to find a straight line that meets the curve at only one point. Here is how this idea can be carried out for the parabola.

4.5.1 Explain why the line $y = 2x - 1$ passes through the point $(1, 1)$ on the parabola $y = x^2$.

4.5.2 The line $y = 2x - 1$ meets the curve $y = x^2$ where $2x - 1 = x^2$. Show that there is only one solution to this equation (which shows that $y = 2x - 1$ is the tangent to the curve $y = x^2$ at $x = 1$).

4.5.3 Similarly show that the line $y = 2ax - a^2$, where a is any number, meets the curve $y = x^2$ at $x = a$.

4.5.4 Show that there is only one solution to the equation $2ax - a^2 = x^2$ (which shows that the line $y = 2ax - a^2$ is the tangent to the curve $y = x^2$ at $x = a$).

4.6   The Slopes of Other Curves

The above method for computing the slope of the parabola seems reasonable, and it can be justified completely by the method of exhaustion. By choosing dx sufficiently small, the value of $\frac{dy}{dx}$ can be made closer to $2x$ than any number except $2x$ itself. Therefore, the only possible value for the slope at P is $2x$. Today we would say that $2x$ is the *limit* of the slope of PQ.

However, seventeenth-century mathematics was less subtle than this. The quotient $\frac{dy}{dx}$ is so convenient to compute that it was taken to *be* the slope at P, even though the quotient $\frac{dy}{dx}$ of infinitesimals is generally not a single number. Whatever the infinitesimal dx may be, $\frac{dx}{2}$ is surely an infinitesimal too, and not equal to dx unless $dx = 0$ (in which case $\frac{dy}{dx}$ has no meaning). Thus the expression for $\frac{dy}{dx}$ computed for the parabola,

$$\frac{dy}{dx} = 2x + dx,$$

is not only ambiguous, it actually *avoids* the value $2x$. It represents a range of values infinitesimally close to $2x$. To obtain the "correct" value $2x$ we apparently need dx to be zero—after previously forcing it to be *non*zero in order to divide by it.

How frustrating! In the seventeenth century various attempts were made to resolve this conflict. Some justified the assumption that $\frac{dy}{dx}$ is the slope of the curve at P by saying, as the Marquis l'Hôpital did in the first textbook of infinitesimal calculus in 1696 [33, p. 2],

> … two quantities that differ by an infinitesimal are the same.

The geometric equivalent of the assumption is to suppose that the infinitesimal arc of the curve between P and Q is the same as the infinitesimal line segment PQ. L'Hôpital believed this too [33, p. 3]:

> … a curved line may be regarded as being composed of infinitely many infinitesimal straight line segments.

Both assumptions seem indefensible, but they were tolerated as long as they gave results that could be checked by the method of exhaustion.

A more defensible approach grants that infinitesimals are smaller than nonzero numbers, but accepts the consequence that a "number plus infinitesimal," such as $2x + dx$, is *not* equal to the number $2x$. Instead, the two are connected by a looser notion than equality that Fermat called *adequality*. If we denote adequality by $=_{\text{ad}}$, then it is accurate to say that

$$2x + dx =_{\text{ad}} 2x,$$

and hence that $\frac{dy}{dx}$ for the parabola is *adequal* to $2x$. Moreover, $2x + dx$ is not a number, so $2x$ is the *only* number to which $\frac{dy}{dx}$ is adequal. This is the true sense in which $\frac{dy}{dx}$ represents the slope of the curve.

Fermat introduced the idea of adequality in the 1630s but he was ahead of his time. His successors were unwilling to give up the convenience of ordinary equations, preferring to use equality loosely rather than to use adequality accurately. The idea of adequality was revived only in the twentieth century, in the so-called *nonstandard analysis* (see Section 4.9.)

For the moment, let us go with the seventeenth-century flow and compute the slope of some curves as Leibniz or l'Hôpital would have done. First, consider the curve $y = x^3$. This has the shape shown in Figure 4.12, but we can calculate the slope at the point (x, y) without reference to the figure.

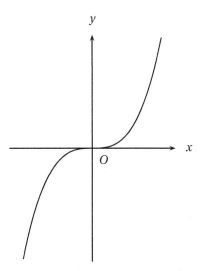

Figure 4.12: The curve $y = x^3$.

Since $y = x^3$, $y + dy = (x + dx)^3$, and hence the height difference dy between points $P = (x, y)$ and $Q = (x + dx, y + dy)$ is

$$dy = (x + dx)^3 - x^3 = x^3 + 3x^2 dx + 3x(dx)^2 + (dx)^3 - x^3$$
$$= 3x^2 dx + 3x(dx)^2 + (dx)^3.$$

Therefore, the slope of the infinitesimal line segment PQ is

$$\frac{dy}{dx} = 3x^2 + 3x\,dx + (dx)^2.$$

Ignoring the infinitesimal dx, we get

$$\text{slope at } P = 3x^2.$$

The Slope of $y = x^{n+1}$

It is easy to generalize this calculation to $y = x^{n+1}$ for any natural number n. Just as we have found

$$(x + dx)^2 = x^2 + 2x\,dx + \text{terms with higher powers of } dx,$$
$$(x + dx)^3 = x^3 + 3x\,dx + \text{terms with higher powers of } dx,$$

we find that

$$(x + dx)^4 = x^4 + 4x^3\,dx + \text{terms with higher powers of } dx,$$
$$(x + dx)^5 = x^5 + 5x^4\,dx + \text{terms with higher powers of } dx,$$

$$\vdots$$

The pattern continues from each exponent n to the next because, if

$$(x + dx)^n = x^n + nx^{n-1}dx + \text{terms with higher powers of } dx,$$

then multiplication gives

$$(x + dx)^{n+1} = (x + dx)(x^n + nx^{n-1}dx + \text{terms with higher powers of } dx)$$
$$= x^{n+1} + (n + 1)x^n\,dx + \text{terms with higher powers of } dx.$$

The height difference dy between $P = (x, y)$ and $Q = (x+dx, y+dy)$ on the curve $y = x^{n+1}$ is therefore given by

$$dy = (x+dx)^{n+1} - x^{n+1} = (n+1)x^n\,dx + \text{terms with higher powers of } dx,$$

and hence

$$\text{slope of } PQ = \frac{dy}{dx} = (n + 1)x^n + \text{terms with powers of } dx.$$

Finally, letting dx vanish, we get the slope of $y = x^{n+1}$ at the point $P = (x, y)$:

$$\text{slope at } P = (n + 1)x^n.$$

Exercises

A tangent to the curve $y = x^3$ can be found without using infinitesimals (as in the previous exercise set), by finding a straight line that meets the curve at only *two* points (one of which it "meets twice"—this is the tangent point).

4.6.1 Explain why the line $y = 3x - 2$ passes through the point $(1, 1)$ on the curve $y = x^3$.

4.6.2 The line $y = 3x - 2$ meets the curve $y = x^3$ where $3x - 2 = x^3$, that is, where $x^3 - 3x + 2 = 0$. Show that $x^3 - 3x + 2 = (x - 1)^2 (x + 2)$, so that $x = 1$ is twice a solution of $x^3 - 3x + 2 = 0$.

4.6.3 Deduce from Question 4.6.2 that the line $y = 3x - 2$ meets the curve at two points, $(-2, -8)$ and $(1, 1)$. (We say that the line "meets the curve twice" at $x = 1$ because $x = 1$ is twice a solution.)

4.6.4 Explain, with the help of a sketch of the curve $y = x^3$, why a line through the points $(1, 1)$ and $(-2, -8)$ that is *not* tangent at $(1, 1)$ will meet the curve at three points.

4.7 Slope and Area

The main reason that infinitesimals refused to die, despite their position on the edge of impossibility, is the powerfully suggestive Leibniz notation. The d notation for infinitesimal differences is only half the story; there is also a notation $\int$ (an elongated S, today known as the integral sign) for *sums.*

Sums of infinitesimals occur as areas and volumes of curved figures. For example, suppose we seek the area beneath the parabola $y = x^2$ between the values $x = 0$ and $x = 1$ (Figure 4.13).

We imagine this area filled with strips of height y and infinitesimal width dx, like the one shown in the figure (only much thinner—we had to draw the strip fairly wide in order to write "dx" inside it). Thus the area under the curve is a sum of areas $y\,dx$ of infinitesimal rectangles.

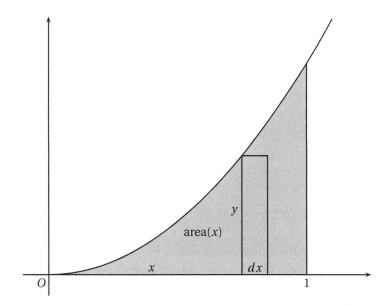

Figure 4.13: The area under the parabola.

Leibniz' symbol for this sum is

$$\int y\,dx.$$

In fact, this is the symbol for the area under any curve. We get a formula for a specific curve by replacing y by the appropriate function of x. In the case of the parabola $y = x^2$, the area is

$$\int x^2\,dx.$$

We also have to specify where the sum begins and ends. In this case it begins at $x = 0$ and ends at $x = 1$, so we write

$$\text{area under parabola between 0 and 1} = \int_0^1 x^2\,dx.$$

To find this area we solve the more general problem of finding the area between 0 and any value of x. We denote this area by area(x), and we find the function area(x) by finding its slope. This is easy to do by applying the infinitesimal difference operation d to area(x), and using equality in the loose Leibnizian manner:

$d\,\text{area}(x) = \text{area}(x + dx) - \text{area}(x)$

$\qquad = x^2 dx \qquad$ because the areas differ by a strip of height $y = x^2$.

Dividing both sides by dx, we get the slope of the area(x) function:

$$\frac{d\,\text{area}(x)}{dx} = x^2.$$

Thus area(x) is a function with value 0 at $x = 0$, and slope x^2 at any value x.

We already know a function which is 0 at $x = 0$, and with slope $3x^2$ at x, namely the function x^3. (We found its slope in the previous section.) And it is clear from Figure 4.14 that tripling the height of any function triples its slope, so $x^3/3$ is a function with the slope x^2 we want.

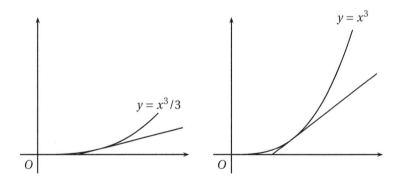

Figure 4.14: Triple the height, triple the slope.

In fact, the function $x^3/3$ exactly equals area(x), because their difference $x^3/3 - \text{area}(x)$ has value 0 at $x = 0$ and slope 0, hence it is always 0.

To summarize: we have area(x) = $x^3/3$. In particular area(1)=1/3, that is

$$\int_0^1 x^2\,dx = \frac{1}{3}.$$

We can similarly use the slope $(n+1)x^n$ of x^{n+1} from the previous section to show that the area under the curve $y = x^n$ is given by the function $x^{n+1}/(n+1)$, and hence that

$$\int_0^1 x^n dx = \frac{1}{n+1}.$$

This turns out to be important in finding a formula for π, as we shall see in the next section.

Figure 4.15 shows the curves $y = x^{n+1}$, and their areas, for $n = 0, 1, 2, 3$. For $n = 0$ the curve is of course a straight line, and for $n = 1$ it is a parabola.

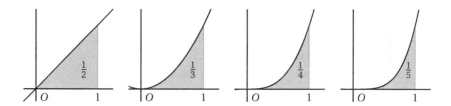

Figure 4.15: The area under $y = x^{n+1}$ for $n = 0, 1, 2, 3$.

The Fundamental Theorem of Calculus

Leibniz's version of the fundamental theorem is simply the "obvious" relation between sums and differences of infinitesimals: $d \int y \, dx = y \, dx$. In words, the "difference" between "successive" sums equals the "last" term in the sum. However, all the words in quotes are based on the fantasy that the area $d \int y \, dx$ is a finite sum of infinitesimal terms $y \, dx$. Actually, there are no "sums" in this sense, hence no "successive" sums and no "last" term.

In rigorous treatments of calculus the proof is more substantial, because it has to deal with genuine sums, though it is nevertheless guided by the infinitesimal fantasy. The theorem is "fundamental" because it reduces the problem of computing areas to the simpler problem of finding slopes.

Exercises

4.7.1 Check directly that the curve $y = x^2/2$ has slope x for any x by calculating

$$\frac{(x + dx)^2/2 - x^2/2}{dx}$$

and setting $dx = 0$ after simplifying the fraction.

4.7.2 Similarly check by calculation that the curve $y = x^3/3$ has slope x^2 for any x.

4.8   The Value of π

In Figure 4.16, the tangent AB of length x corresponds to an arc of the unit circle whose length is called arctan x. If we increase x by an infinitesimal amount dx, then $y = \arctan x$ increases by an infinitesimal amount dy. We now use the figure to prove that $dy = \frac{dx}{1+x^2}$.

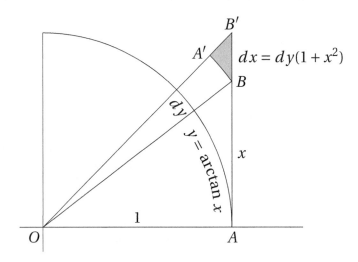

Figure 4.16: Infinitesimal properties of the arc tangent.

Since dy lies on a circle of radius 1, the corresponding arc BA' on the circle with radius OB has length $dy\sqrt{1 + x^2}$, since $OB = \sqrt{1 + x^2}$ by Pythagoras' theorem in triangle OAB.

Since the arc BA' is infinitesimal, we can view $B'A'B$ as an infinitesimal triangle (shaded), similar to triangle BAO because the angle between the lines OB and OA' is infinitesimal. It follows that the infinitesimal hypotenuse $dx = BB'$ is $\sqrt{1+x^2}$ times the length of the side $BA' = dy\sqrt{1+x^2}$. This explains the value shown, $dx = dy(\sqrt{1+x^2})^2 = dy(1+x^2)$, and gives us the required value of dy,

$$dy = \frac{dx}{1+x^2}.$$

Summing all the infinitesimals dy for x from 0 to 1 we find that

$$\arctan 1 = \int_0^1 \frac{dx}{1+x^2}.$$

Now we need only geometric series and the fundamental theorem of calculus to prove the astonishing result that

$$\frac{\pi}{4} = 1 - \frac{1}{3} + \frac{1}{5} - \frac{1}{7} + \frac{1}{9} - \cdots.$$

The proof goes as follows:

$$\frac{\pi}{4} = \arctan 1 = \int_0^1 \frac{dx}{1+x^2}$$

$$= \int_0^1 (1 - x^2 + x^4 - x^6 + x^8 - \cdots)\,dx \qquad \text{by geometric series}$$

$$= 1 - \frac{1}{3} + \frac{1}{5} - \frac{1}{7} + \frac{1}{9} - \cdots \qquad \text{since } \int_0^1 x^n\,dx = \frac{1}{n+1}.$$

This result was first discovered by Indian mathematicians around 1500 CE and rediscovered in Europe by the more general method that uses the fundamental theorem of calculus. Other than that, the two discoveries follow a similar pattern, involving the formula

$$d\arctan x = \frac{dx}{1+x^2}$$

and the geometric series

$$\frac{1}{1+x^2} = 1 - x^2 + x^4 - x^6 + x^8 - \cdots.$$

The result is wonderful in itself—who would have thought that π is encoded by the sequence of odd numbers?—but even more astonishing is its independent appearance in two different civilizations. It suggests not only that this is the simplest and most natural formula for π but also that π has as much to do with the natural numbers as with geometry.

4.9   Ghosts of Departed Quantities

> ... she waited for a few minutes to see if she was going to shrink any further: she felt a little nervous about this; "for it might end, you know," said Alice to herself, "in my going out altogether, like a candle. I wonder what I should be like then?" And she tried to fancy what the flame of a candle looks like after the candle is blown out, for she could not remember having seen such a thing.

> Lewis Carroll, *Alice's Adventures in Wonderland*, Chapter 1

Infinitesimals have a Wonderland quality because we want them to be like tiny Alices that end by going out altogether. But is it possible for mathematical objects to behave in this way? The paradoxical, even inconsistent, behavior that mathematicians attributed to infinitesimals was criticized early and often by philosophers. In [28, p. 301] we find Thomas Hobbes in 1656, attacking the "indivisibles" (infinitesimal slices of solid bodies used to determine volumes) used by the Oxford mathematics professor John Wallis:

> Your scurvy book of *Arithmetica infinitorum*; where your indivisibles have nothing to do, but as they are supposed to have quantity, that is to say, to be *divisibles*.

Hobbes was right (if not exactly polite) to object to the way infinitesimals were used, but he objected to the *results* of infinitesimal calculus as well. In 1672 he ridiculed the discovery of a solid body with finite volume and infinite surface area [28, p. 445]:

> ... to understand this for sense, it is not required that a man should be a geometrician or a logician, but that he should be mad.

He did this because he thought he knew better, in fact he believed he had solved a problem that had defeated the best mathematicians. The problem was none other than "squaring the circle," or in our terms, finding the value of π. Hobbes "solved" the problem essentially by abolishing the circle, declaring that points are physical objects, and hence they have size greater than zero (for more details, see the book *Squaring the Circle* by Douglas Jesseph). This bizarre and pathetic episode made Hobbes a laughing-stock in the eyes of mathematicians, probably just increasing their complacency about infinitesimals. After all, Leibniz was a great philosopher, and *he* was on their side.

Philosophy struck back in 1734, when Bishop George Berkeley wrote the first really effective criticism of infinitesimal calculus, pointing out inconsistencies in the work of Leibniz, Newton, and their followers with great humor and vigor. Berkeley did not question the results of calculus, in fact he regarded them as provable by more rigorous means. But in his *Analyst* [4, section 35] he mocked the supernatural behavior of infinitesimals, or "evanescent increments" as Newton called them:

> Whatever is got ... is to be ascribed to fluxions: which must therefore be previously understood. And what are these fluxions? The velocities of evanescent increments? And what are these same evanescent increments? They are neither finite quantities, nor quantities infinitely small yet nothing. May we not call them ghosts of departed quantities?

Berkeley's criticism stung, and mathematicians tried to answer it, though for a long time without much success. The problem was deeper than anybody realized, being entangled with the concepts of number and infinity. As we have already seen, mathematicians were wrestling with the problem of irrational numbers until the late nineteenth century, and we have barely begun to see all the problems of infinity (for more, see Chapter 9). Nevertheless, between 1830 and 1870 a serviceable approach to calculus was worked out, based on the concepts of function and limit.

This is the mainstream approach to calculus used today. It denies the existence of infinitesimals, and interprets the word "infinitesimal" as a mere figure of speech in statements that are properly made using limits. For example "let dx be infinitesimal" would be restated as "let Δx tend to zero." However, even the mainstream approach uses the Leibniz notations $\frac{dy}{dx}$ and $\int y\,dx$, because they are so concise and suggestive.

This leads to some awkward moments. It has to be explained that $\frac{dy}{dx}$ is *not* the ratio of infinitesimal differences dy and dx—since infinitesimals do not exist—but is rather a symbol for the *limit of a ratio* $\frac{\Delta y}{\Delta x}$ as Δx tends to zero, where Δx is a finite change in x and Δy is the corresponding change in the function y of x. Likewise, $\int y\,dx$ is not an actual sum of terms $y\,dx$, but the *limit* of a sum of terms $y\Delta x$. Thus avoidance of infinitesimals comes at the cost of a strange dual notation: Δ for actual differences and d (the ghost of Leibniz!) for the limits of their quotients and sums.

To many this is a compromise solution which fails to explain why infinitesimals work. Is it possible to define and use genuine infinitesimals?

Since an infinitesimal is required to be smaller than any nonzero number, but not zero, infinitesimals are not numbers. They can be *functions of time* however, and this seems in the right spirit. For example, the function $f(t) = 1/t$ makes a good infinitesimal, because it tends to zero and hence is smaller than any given positive number for all sufficiently large t. Functions are like numbers, inasmuch as they can be added, subtracted, multiplied, and divided. And some functions behave exactly like numbers, namely the *constant* functions, which have the same value at all times t. Thus the world of functions is one with objects that behave like numbers (the constant functions) and also objects that behave like infinitesimals (the functions that tend to zero as t tends to infinity).

This larger world resolves the paradox of infinitesimals, though we have glossed over one problem. We also need infinitesimals to be *ordered*. That is, if $g(t)$ and $h(t)$ are two functions that tend to zero as t tends to infinity, we want one of them to be "less" than the other. For example, we have to decide which is the larger of $g(t) = \frac{\sin t}{t}$ and $h(t) = \frac{\cos t}{t}$ (Figure 4.17). It is hard to make all such decisions consistently, though it can be done through advanced methods of logic.

The first to solve this problem completely was the American mathematician Abraham Robinson, in the 1960s. His system is called *nonstandard analysis*, and it has been successful enough to yield some new

Figure 4.17: Which function is larger?

results. However, nonstandard analysis is not yet as simple as the old
Leibniz calculus of infinitesimals, and there is a continuing search for
a really natural system that uses infinitesimals in a consistent way.

Chapter 5

Curved Space

Preview

People of ancient and medieval times are often said to have believed that the earth was flat, a belief supposedly overthrown by Christopher Columbus. This is a myth. *Not only did the ancients know that the earth was round, they believed that* space *was round too—an idea that seems impossible to most people today*. We are used to thinking of roundness as a property of objects *in* space, such as circles or spheres, but not *of* space itself. *Evidently our modern spatial intuition needs more experience with "impossible" forms of space*.

In the present chapter we explore ways of visualizing space, starting with the mathematically simplest case—the infinite space of Euclid's geometry. It is known as *Euclidean space* and it is called *flat* because of its analogy with Euclid's flat plane. Next we look at *spherical space*, a finite "round" space that Dante seems to describe in his *Divine Comedy*. We describe it mathematically by analogy with the spherical surface. Finally we look at *hyperbolic space*, an infinite space even more "spacious" than Euclidean space. It contains *hyperbolic planes*—surfaces more spacious than Euclidean planes. In a hyperbolic plane there is more than one parallel to a given line through a given point.

Spherical space and hyperbolic space have *curvature* analogous to the curvature of surfaces. It can be detected by the behavior of lines, and in particular by whether or not parallels exist and are unique. Curvature also gives the answer to an ancient question in geometry: is

134

Euclid's parallel axiom a consequence of his other axioms? The existence of hyperbolic space enables us to say *no*, and thus *there is such
a thing as* non-Euclidean geometry—*something considered impossible
until quite recently.*

5.1   Flat Space and Medieval Space

> Assume though, for a moment, that all space
> Is definitely limited, what happens
> If somebody runs to its furthest rim, and rifles
> A javelin outward? Will it keep on going,
> Full force, or do you think something can stop it?
> Here's a dilemma you can't escape!
> You have to grant an infinite universe
> For either there's matter there to stop your spear,
> Or space through which it keeps on flying. Right?
>
> Rolfe Humphries' translation of Book I, lines 970–978, of
> Lucretius' *De rerum natura* (The title is usually translated
> as *On the nature of things*, but Humphries translates it as
> *The way things are.*)

Lucretius gives apt and vivid expression to an idea that must have
occurred to almost everybody at some time: space must be infinite,
because it surely does not come to an end. He probably would have
agreed with the picture of space shown in Figure 5.1, which shows the
geometric structure of the presumed infinite space of our imagination.
Space is three-dimensional, infinite, and *flat*. We say more about flatness later, but for the moment take it to mean that space can be filled
with flat square objects, namely cubes.

Natural though it is, infinite flat space conflicts with ancient cosmology, and Lucretius was trying to debunk a common belief. The
Greeks believed the universe should reflect the geometric perfection
of circles and spheres, and they imagined space structured by a system
of spheres. Their mental picture is shown (in cross section) in the left
half of Figure 5.2. The earth is the innermost sphere, surrounded by
eight concentric "heavenly" spheres carrying the known celestial bodies, and an outermost sphere called the Primum Mobile. (For exam-

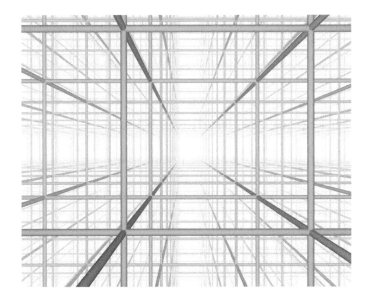

Figure 5.1: Infinite flat space.

ple, "seventh heaven" is the sphere of Saturn.) The motion of the sun, moon, planets, and stars was attributed to the rotation of the spheres carrying them, with the Primum Mobile ("first mover") controlling them all. Somehow, the ancient universe stopped at the Primum Mobile, an idea that Lucretius skewered with his javelin.

Medieval theology filled in the blank beyond the Primum Mobile with the *Empyrean* (right half of Figure 5.2)—the home of God and the angels—which is a structure of concentric "angelic" spheres outside the Primum Mobile. God is a light at the center of the innermost sphere in the Empyrean, in some way opposite to Satan, who is at the center of the earth.

This is an elegant, conceptual way to fill the blank, but it is hard to fit the two halves of the medieval universe smoothly together. Describing this universe plausibly is a greater poetic challenge than describing infinite flat space; in fact, it might be considered impossible. However, the Middle Ages found a poet equal to the task: the great Dante Alighieri (1265–1321). Dante's work the *Divine Comedy* is best known for the part about hell, the *Inferno*, but its third part, the *Paradiso*, is fascinating from the viewpoint of geometry and astronomy. In

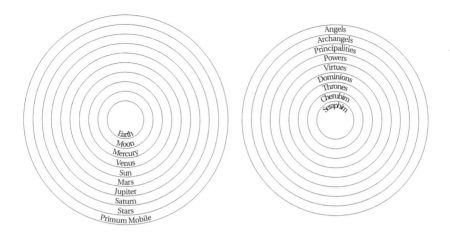

Figure 5.2: The heavenly and angelic spheres.

Canto XXVIII Dante views the Empyrean as not only the *complement* but also the *reflection* of the heavens visible from earth. He makes a smooth transition from the heavens to the Empyrean by using the Primum Mobile as a half-way stage between two worlds, the "model" and the "copy." From this vantage point, he sees the heavenly spheres on one side as an image of the angelic spheres on the other,

> as one who in a mirror catches sight
> of candlelight aglow behind his back
> before he sees it or expects it,
>
> and, turning from the looking-glass to test
> the truth of it, he sees that glass and flame
> are in accord as notes to music's beat
>
> (Translation by Mark Musa of lines 4–9, Canto XVIII, of Dante's
> *Paradiso.*)

With this sophisticated model of a finite universe, the Church was able to hold out against infinite space for a few centuries. But eventually infinite flat space came to be generally accepted for its greater simplicity, despite some uneasiness about infinity we will explore further in Chapter 9.

In the twentieth century, cosmology returned to the idea of a finite universe, and physicists now look back in admiration to Dante's *Par-*

adiso, seeing in it a good description of the simplest finite universe, which we now call the *3-sphere*. The first such homage to Dante that I know is the paper "Dante and the 3-sphere" by Mark Peterson in the *American Journal of Physics* 47 (1979), pages 1031–1035. Peterson offers a few different ways to think of Dante's universe, mainly from a qualitative point of view. In the next section I offer a more quantitive approach, mapping the 3-sphere as faithfully as possible into "ordinary" flat space.

5.2 The 2-Sphere and the 3-Sphere

To understand the 3-sphere in the context of flat three-dimensional space it helps to look at the ordinary spherical surface, which mathematicians call the *2-sphere*. We consider it from the viewpoint of creatures who believe they live in the plane—call them *flatearthers*. How might they experience the 2-sphere, and how might they represent their experience in a planar diagram?

Suppose a flatearther is placed at the north pole N of a 2-sphere, and that he explores this world by drawing larger and larger circles with center N (Figure 5.3). These are what we call *latitude circles*. At first (left picture in Figure 5.3) these concentric circles will have noticeable curvature, and it will be clear which side is "inside" and which is "outside." But at the equator (middle picture) the circle will seem straight, and beyond this (right picture) "inside" and "outside" change places, and the south pole becomes the center.

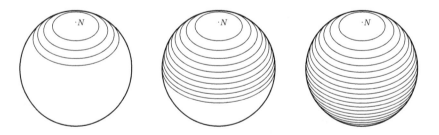

Figure 5.3: Latitude circles on the 2-sphere.

The appearance of a point "opposite" to his starting point at the north pole will come as a great surprise to the flatearther, just as a point in space opposite to the center of the earth seems strange to us. (We are "flatspacers.") However, the flatearther can capture his experience fairly faithfully by a planar diagram such as Figure 5.4, in which circles expand outwards from a point N, decreasing in curvature until they become straight, then bend the other way and shrink in towards another point S.

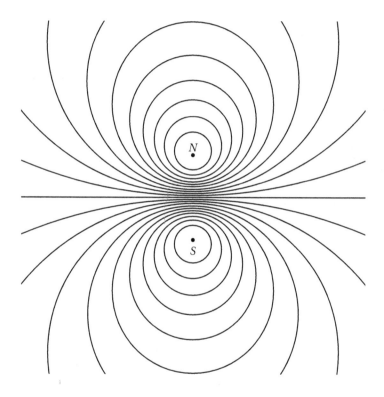

Figure 5.4: Planar image of the 2-sphere.

In fact, if you put your eye very close to the paper at the center of Figure 5.4 (or, better, near the center of a blown-up photocopy) you will actually see what the 2-sphere looks like *from the inside* when viewed from a point on the equator. This is because Figure 5.4 was obtained by the process called *stereographic projection*, shown in Figure 5.5.

Stereographic projection goes back to Claudius Ptolemy around

150 CE, and possibly to Hipparchus 300 years earlier. The sphere is projected from one of its points, onto a plane touching the sphere at the opposite point.

If the sphere is made of glass, with latitude circles painted on it, then a light at the point of projection will cast shadows on the plane in the pattern shown in Figure 5.4 (here we project from a point on the equator, so N and S are the shadows of the poles). If the eye is placed at the point of projection, as in Figure 5.5, its view of the pattern is exactly the same as its view of the inside of the sphere.

Projection of course does not give a completely faithful image of the sphere. For example, it maps a finite circle (the equator) to an infinite line. But stereographic projection maps an infinitesimal figure on the sphere to an infinitesimal figure of the same shape on the plane, so it is "faithful in the small." This property, highly desirable for maps of the earth, was discovered by the English mathematician (and assistant to the explorer Sir Walter Raleigh) Thomas Harriot around 1590. Stereographic projection maps circles to circles (or, in exceptional cases, to straight lines), so it also preserves the shape of certain large figures. Its circle-preserving property was already known to Ptolemy.

How does all this help us with the 3-sphere? Well, *there is a stereographic image of the 3-sphere in ordinary three-dimensional space.* It has properties analogous to those of the image of the 2-sphere on the plane.

- A sequence of concentric "latitude 2-spheres" in the 3-sphere is mapped to a nested sequence of 2-spheres in space.

- The image spheres initially expand out from the starting point N, then seem to bend the other way and shrink down to a second point S—the "opposite pole" to N.

To construct this stereographic image of the 3-sphere we simply take each of the circles in Figure 5.4 as the equator of a 2-sphere.

I believe this is easy enough to imagine if you look again at Figure 5.4, so instead I would like to show a more artistic version of the idea. Pictures showing both the earth and paradise are common in Italian Renaissance art, and they often try to show the opposed heavenly and angelic spheres. Indeed, the idea is so common that it also appears in pictures with secular themes. My favorite is Tintoretto's *Allegory of*

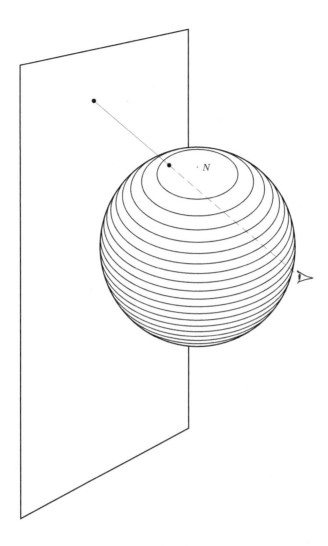

Figure 5.5: Stereographic projection of the 2-sphere.

Venice as Queen of the Sea, shown in Figure 5.6. (I got the idea of looking at Tintoretto from the 1925 book *Klassische Stücke der Mathematik*
by Andreas Speiser. At the beginning of the book Speiser shows another
painting of Tintoretto and comments on its resemblance to the interior
of a 2-sphere with latitude circles drawn on it. I think that the Venice
picture looks more like a 3-sphere.)

Figure 5.6: Tintoretto's *Allegory of Venice as Queen of the Sea.*

Exercises

On the sphere we can talk about *lines, angles,* and *triangles.* A "line" is a *great circle,* which is the curve cut into the sphere by a plane through its center. This is a "line" in the sense that it gives the shortest distance (on the sphere) between any two of its points, and we discuss such "lines" further in Sections 5.3 and 5.4. The "angle" between any "lines" is the angle between the corresponding planes. A *spherical triangle* is formed by three distinct lines on the sphere, and there are some interesting *tilings* by identical spherical triangles. Figure 5.7 shows an example.

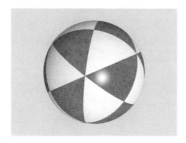

Figure 5.7: Tiling the sphere by triangles.

5.2.1 Explain, by examining the angles that meet at each vertex, why each triangle has angles $\pi/2$, $\pi/3$, and $\pi/3$. (Hint: the angles at each vertex are equal, and they have to add up to 2π, so the angles can be found simply by counting.)

5.2.2 Similarly find the angles of each triangle in the two tilings shown in Figure 5.8.

Figure 5.9 shows a sphere divided into the same spherical triangles as in the first example in Figure 5.8, but with every other triangle cut out.

5.2.3 How many triangles are there? (Hint: Focus on a piece of the sphere bounded by three great circles that meet at right angles.)

Every other triangle on the sphere has been cut out so that a light inside the north pole will cast a shadow on the plane—giving the stereographic projection of the triangle pattern shown in Figure 5.9.

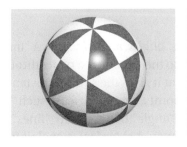

Figure 5.8: Other tilings of the sphere by triangles.

Figure 5.9: Sphere with cutout triangles.

Figure 5.10: Stereographic projection of spherical triangles.

5.2.4 Use the stereographic projection to check the number of triangles on the sphere.

5.3 Flat Surfaces and the Parallel Axiom

The 3-sphere is an example of a *curved space*, an idea that can now be grasped through its analogy with the 2-sphere, which is clearly a curved surface. We have always thought 2-spheres are curved because we see them from the "outside," where the curvature is clearly visible. But now we also know how curvature can be detected from *within* the surface—an idea that can be extended to detect curvature of space, which we *cannot* view from "outside." The idea of studying a surface "internally," or *intrinsically*, was first pursued systematically by Gauss, though of course the sphere had been explored intrinsically much earlier in connection with astronomy, navigation, and surveying. Gauss himself was responsible for surveying the Kingdom of Hanover in the 1820s, so he may have been generalizing his own experience with the earth's surface.

The first task in developing the intrinsic geometry of surfaces is to identify the "straight lines." On any smooth surface it makes sense to call a curve segment "straight," or *geodesic*, if it is the shortest path between its endpoints. Between any two sufficiently close points A and B there is in fact a *unique* geodesic segment AB, which may be found experimentally by stretching a thread tightly over the surface from A to B. A "line," or *geodesic*, may then be defined as a curve whose sufficiently short segments are all geodesic segments.

On the plane, the geodesics are ordinary lines, and on the 2-sphere they are the *great circles* (the intersections of the 2-sphere with planes through its center). We shall further investigate the geometry of great circles, and how they reveal the curvature of the sphere, in the next section. For the moment, we wish to study geodesics on another *flat* surface, the cylinder, in order to see some properties of geodesics that might be wrongly attributed to curvature. The cylinder looks curved, but it is not *intrinsically* curved.

The Cylinder

The cylinder is an intrinsically flat surface because it is constructed by rolling up the plane, which is absolutely flat. The fact that the plane is bent in order to roll it up does *not* make the cylinder intrinsically curved. Creatures living in it would not observe any difference between small regions of the cylinder and small regions of the plane. They would consider it "locally flat," and intrinsic flatness means no more than this. The concept of surface curvature is supposed to measure how much the surface deviates from the plane *in the neighborhood of a point*, and hence it is all about behavior in small regions.

So the cylinder is flat, but of course it is not the same as the plane. Its differences become clear over large regions, particularly in the behavior of geodesics. A short geodesic segment behaves the same as a short line segment in the plane, but a geodesic is a "rolled up line," and it can take three quite different forms, shown in Figure 5.11.

Figure 5.11: Geodesics on the cylinder.

In the orientation we have given the cylinder in Figure 5.11, the geodesics are

- horizontal straight lines,

- helices, which spiral around the cylinder, and

- vertical circles.

Thus the geodesics on the cylinder behave differently from straight lines in the plane in at least two respects:

- Some of them are finite (the circles).

- There may be two or more geodesics passing through the same two points (for example, a helix and a horizontal straight line).

The latter two properties of geodesics are therefore *not* an indication that the surface is intrinsically curved. To detect curvature we need to find deviations from the geometry of the plane that occur in small regions.

We have seen in Chapter 3 how plane geometry is ruled by the parallel axiom, so "small-scale" consequences of the parallel axiom hold in small regions of the plane and other flat surfaces. One such consequence is the Pythagorean theorem. It holds for arbitrarily small triangles, hence it holds for sufficiently small triangles on flat surfaces such as the cylinder. One way to detect curvature (of the sphere, say) would therefore be to show that the Pythagorean theorem fails. This is possible, but it is more elegant to use another small-scale consequence of the parallel axiom: the theorem on the angle sum of a triangle. (We introduced this theorem as an exercise in Section 3.1.)

The Angle Sum Theorem

The theorem states that *the angles of a triangle sum to two right angles.* One can see how it follows from the parallel axiom with the help of two diagrams (Figure 5.12).

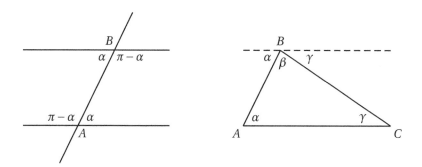

Figure 5.12: Angles governed by the parallel axiom.

The diagram on the left shows a pair of parallels crossed by a line AB. Since parallels do not meet, the angles AB makes with the parallels, on either side, cannot sum to less than two right angles (by Euclid's version of the parallel axiom, explained in Section 3.1). The only possibility is that on each side the angles sum to precisely two right angles, that is, to π. So AB makes the angles α and $\pi - \alpha$ as shown.

Now consider the diagram on the right: a triangle ABC, with angles α, β, and γ as shown, and with the parallel to AC through B added. By what we have just seen from the diagram on the left, there is another copy of the angle α at B as shown, and similarly another copy of γ. But then $\alpha + \beta + \gamma$ also represents the straight angle at B, and therefore $\alpha + \beta + \gamma = \pi$. $\square$

(Remember, the sign $\square$ signals the end of a proof.)

In the next section we shall see that the angle sum of triangles can detect curvature, by showing that the angle-sum theorem fails in a very interesting way on the 2-sphere.

Exercises

Small regions of the cylinder look exactly like regions of the plane, but large regions may behave differently. In particular, "large" geodesics on the cylinder may not behave like large straight lines in the plane.

5.3.1 Give examples of geodesics on the cylinder that

- do not meet,
- meet in only one point,
- meet in infinitely many points.

5.3.2 Also give an example of two points that are connected by only one geodesic.

5.4   The Sphere and the Parallel Axiom

As mentioned in Section 5.3, the geodesics on the 2-sphere are great circles: the intersections of the 2-sphere with planes through its center. Figure 5.13 shows some of them, which form a *spherical triangle ABC*.

Geodesics on the sphere behave differently from straight lines in the plane in at least three respects:

- They are finite closed curves.

- Certain pairs of points (for example the north and south pole) are connected by more than one geodesic.

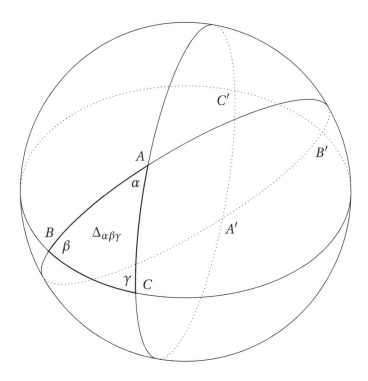

Figure 5.13: Great circles and a spherical triangle.

- There are no parallels; in fact, any two geodesics meet in two points (diametrically opposite to each other).

From our experience with the cylinder we know that the first two of these properties are not a sure sign of surface curvature. However, the third property *is* such a sign. It implies that *in every spherical triangle the angle sum is greater than* π. Hence no region of the 2-sphere, no matter how small, behaves like a region of the plane.

In fact, not only is $\alpha + \beta + \gamma > \pi$ for any spherical triangle with angles α, β, and γ; *the difference* $\alpha + \beta + \gamma - \pi$ *is proportional to the area* $\Delta_{\alpha\beta\gamma}$ *of the triangle*. This beautiful theorem—the key to *spherical geometry*—was discovered by Thomas Harriot in 1603 and later extended by Gauss to a general connection between surface curvature and angle sums of triangles.

Harriot proved his theorem by considering a *slice* of the sphere cut out by two great circles (Figure 5.14). Two planes through the center of the sphere cut a pair of identical slices from its surface, each with area proportional to the angle α between the planes. Thus if S is the total surface area of the sphere, the area of a slice with angle α is $\frac{\alpha}{2\pi}S$.

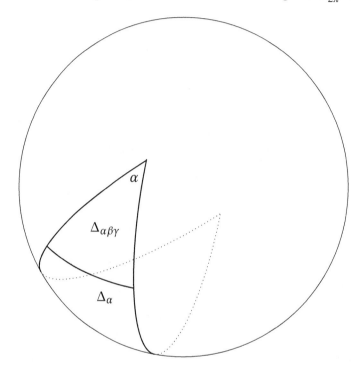

Figure 5.14: Area of a slice.

Now the area $\Delta_{\alpha\beta\gamma}$, together with the adjacent triangular area Δ_α, makes a slice of angle α, hence

$$\Delta_{\alpha\beta\gamma} + \Delta_\alpha = \frac{\alpha}{2\pi}S.$$

Similarly,

$$\Delta_{\alpha\beta\gamma} + \Delta_\beta = \frac{\beta}{2\pi}S$$

and

$$\Delta_{\alpha\beta\gamma} + \Delta_\gamma = \frac{\gamma}{2\pi}S,$$

where Δ_β and Δ_γ are the areas of the spherical triangles in Figure 5.13 adjacent to the sides CA and AB, respectively. Adding these three equations we get

$$3\Delta_{\alpha\beta\gamma} + \Delta_\alpha + \Delta_\beta + \Delta_\gamma = \frac{\alpha + \beta + \gamma}{2\pi} S. \tag{1}$$

On the other hand, adding all the eight triangles that make up the surface of the sphere (Figure 5.13) gives

$$2\Delta_{\alpha\beta\gamma} + 2\Delta_\alpha + 2\Delta_\beta + 2\Delta_\gamma = S,$$

and therefore

$$\Delta_{\alpha\beta\gamma} + \Delta_\alpha + \Delta_\beta + \Delta_\gamma = \frac{S}{2}. \tag{2}$$

Subtracting Equation (2) from Equation (1) gives

$$2\Delta_{\alpha\beta\gamma} = \left(\frac{\alpha + \beta + \gamma}{2\pi} - \frac{1}{2}\right)S = \frac{\alpha + \beta + \gamma - \pi}{2\pi} S,$$

so finally

$$\Delta_{\alpha\beta\gamma} = (\alpha + \beta + \gamma - \pi)\frac{S}{4\pi}.$$

Thus the area $\Delta_{\alpha\beta\gamma}$ is proportional to $\alpha + \beta + \gamma - \pi$, as claimed. $\qquad \square$

With the help of calculus it can be shown that the surface area S of a sphere of radius R is $4\pi R^2$, hence in fact

$$\frac{\Delta_{\alpha\beta\gamma}}{\alpha + \beta + \gamma - \pi} = R^2.$$

It is reasonable to define the curvature of a circle of radius R to be $1/R$ (the larger the radius, the smaller the curvature), and the curvature of a sphere of radius R to be $1/R^2$. $1/R^2$ is called the *Gaussian* curvature of the sphere. We discuss Gaussian curvature of other surfaces in Section 5.6.

The above formula for R^2 shows that the Gaussian curvature of a sphere may be found from measurements *within* the surface: the area and angle sum of a triangle. It also shows how *deviation from flatness is revealed by deviation of the angle sum from π.*

Exercises

In these exercises we assume a sphere of radius $R = 1$, so that its area is 4π, and the area of a spherical triangle with angles α, β, γ is $\alpha + \beta + \gamma - \pi$.

5.4.1 Figure 5.15 shows the sphere divided into eight triangular "octants," each of which has three right angles and is clearly 1/8 of the area of the sphere. Does this agree with the formula for the area of a spherical triangle?

5.4.2 Now check whether your answer to Exercise 5.2.3 agrees with the formula for area, by showing that the formula gives each triangle an area of 1/48 of the area of the sphere. Also show that the angles found in the second example in Exercise 5.2.2 give a triangle whose area is 1/120 of the area of the sphere.

5.4.3 Show, by dividing into triangles, that a spherical quadrilateral with angles $\alpha, \beta, \gamma, \delta$ has area $\alpha + \beta + \gamma + \delta - 2\pi$. More generally, give an argument that the area of any spherical polygon with n sides (an "n-gon") is its angle sum minus $(n-2)\pi$.

Suppose that the surface of a sphere is divided into F "faces" by E great circle "edges," which come together at V "vertices." For example, in Figure 5.15 we have $F = 8$, $E = 12$, and $V = 8$. Now consider all the

Figure 5.15: Division of the sphere by great circle arcs.

angles in all the faces of such a subdivision of the sphere. We calculate the sum of these angles in two different ways.

5.4.4 By collecting the angles around each vertex of the subdivision, show that the sum of all the angles is $2\pi V$. (In the figure, this sum is $2\pi \times 6$.)

5.4.5 By Exercise 5.4.3, the angle sum for each n-gon face is its area plus $(n-2)\pi$, so

$$\text{angle sum of an } n\text{-gon face} = \text{area of face} + n\pi - 2\pi. \qquad (*)$$

When we sum the values of n over all the faces we get $2E$. Why?

5.4.6 Conclude, by summing the equations (*) over all the F faces, that

$$2\pi V = \text{sum of all angles}$$
$$= \text{area of sphere} + 2E\pi - 2F\pi,$$
$$= 4\pi + 2E\pi - 2F\pi,$$

and hence show that $V - E + F = 2$.

Thus we have used the concept of angle to draw a conclusion that is not about angles at all: it is about the *numbers V, E, F* of vertices, edges, and faces. The formula $V - E + F = 2$ is called the *Euler polyhedron formula*, and in fact it holds even when edges are not arcs of great circles—they can be any continuous arcs that do not intersect themselves or each other.

5.5 Non-Euclidean Geometry

Euclid explicitly assumed three properties of straight lines in the plane, which are the main characteristics of what we now call *Euclidean plane geometry*:

- Lines are infinite.

- There is a unique line through any two points.

- There is a unique parallel to any line through a point outside it.

When "lines" are understood as geodesics on a surface, such as the cylinder or the sphere, we have seen that all of these properties can fail. Nevertheless, spherical and cylindrical geometry are not regarded as "non-Euclidean"—or "impossible"—since spheres and cylinders coexist with Euclidean planes in the familiar three-dimensional space, shown in the picture at the beginning of this chapter. The really interesting question, which puzzled mathematicians for 2,000 years, is whether existence and uniqueness of parallels follows from the *other* properties of lines in Euclidean plane geometry.

A truly "non-Euclidean" geometry, if it exists, should have all lines infinite and a unique line through any two points *but* the parallel property should fail in some way. Up until the nineteenth century, no "non-Euclidean" geometry in this sense was ever found, and indeed determined attempts were made to prove such a geometry is impossible.

The most complete attempt was made by the Italian Jesuit Girolamo Saccheri in his 1733 book *Euclides ab omni naevo vindicatus* (Euclid cleared of every flaw). Saccheri's idea was to prove the existence and uniqueness of parallels by showing that any alternative leads to a contradiction. The first alternative is that there are no parallels at all, and Saccheri correctly showed that this contradicts the assumption that all lines are infinite. The other alternative is that *more than one* parallel exists, and in this case a contradiction is harder to find. Saccheri was able to show that if there is more than one parallel to a line *l* through a point *P* then there is a nearest parallel *m*, called an *asymptotic* line, on each side of *P* (Figure 5.16).

Figure 5.16: Asymptotic lines.

Moreover, each asymptotic line has a *common perpendicular at infinity* with the line *l*. This sounds bad, but it is not actually a contradiction (nor is it valid to talk about objects at infinity in Euclid's geometry). Saccheri could only reject it on the grounds that "it is repugnant to the nature of straight lines." In fact, Saccheri had taken the first steps into non-Euclidean geometry, and later mathematicians were to find these properties of non-Euclidean lines not repugnant, but rather attractive. Sometime between 1733 and 1800, *the plausible, but actually impossible, dream of proving Euclid's parallel axiom flipped to the implausible, but actually possible, dream of* accepting *non-Euclidean worlds.*

Gauss was probably the first to take non-Euclidean geometry seriously. Reminiscing later in life, he claimed to have begun thinking about it in his teens. However, he feared ridicule from other mathematicians and did not publish his results, then or later. The first publications appeared in the 1820s: some minor ones by friends of Gauss, and fully worked-out, independent discoveries of the subject by Nikolai Lobachevsky in Russia in 1829 and János Bolyai in Hungary in 1832. Bolyai and Lobachevsky believed they had discovered a new world, so complete and beautiful they were sure it must exist, even though they had no concrete model for it. The following are among the properties of the non-Euclidean world they discovered (or rediscovered):

- The angle sum of a triangle is less than π, and the area of a triangle with angles α, β, γ is proportional to $\pi - \alpha - \beta - \gamma$.

- Non-Euclidean space contains surfaces, known as *horospheres* or "spheres with center at infinity," on which Euclidean plane geometry holds. Also, spherical geometry holds on finite spheres in non-Euclidean space. Thus non-Euclidean geometry, if valid, is consistent with Euclidean and spherical geometry because it contains them both.

- Each basic formula of spherical geometry has a counterpart in non-Euclidean geometry, but with the sine and cosine functions replaced by the so-called *hyperbolic* sine and cosine:

$$\sinh x = -i\sin ix = \frac{e^x - e^{-x}}{2}$$
$$\cosh x = \cos ix = \frac{e^x + e^{-x}}{2}.$$

For example, the circumference of a circle of radius r on the sphere of radius R is $2\pi R \sin(r/R)$. The circumference of a non-Euclidean circle of radius r is $2\pi R \sinh(r/R)$, for some constant R.

The existence of these parallel universes of formulas—spherical and hyperbolic—convinced Lobachevsky that the hyperbolic formulas describe something real, but what? It should be a surface with properties somehow "opposite" to those of the sphere—a *non-Euclidean plane,* later called the *hyperbolic* plane—but it was not until 1868 that such a surface was found. To understand the difficulty in realizing this *hyperbolic geometry*, we have to look more closely at the geometry of surfaces, and particularly at the concept of *negative curvature.*

Exercises

5.5.1 Use the formula $e^{ix} = \cos x + i \sin x$ from Section 2.7 to show that

$$\cos x = \frac{e^{ix} + e^{-ix}}{2}, \quad \sin x = \frac{e^{ix} - e^{-ix}}{2i}.$$

Hence explain the formulas given above for $-i \sin ix$ and $\cos ix$.

Figure 5.17 shows a circle of radius 1, and an angle r in a right-angle triangle. The height of the triangle is therefore $\sin r$.

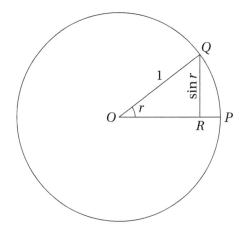

Figure 5.17: Distance, arc length, and angle.

5.5.2 Explain why the arc length from P to Q is r.

We now view Figure 5.17 as a vertical cross-section of a sphere of radius 1, and take the arc PQ as the "radius" of a circle with center P drawn on the sphere (Figure 5.18). It is the circle traced on the sphere by the free end of a string of length r whose other end is fixed at P.

5.5.3 Observe that this circle is in fact the boundary of a vertical disk of radius $\sin r$, and hence it has circumference $2\pi \sin r$.

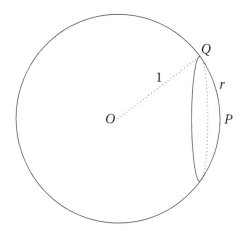

Figure 5.18: Circle of radius r with center P on the sphere.

5.5.4 By extending the argument just given for a circle on the sphere
of radius 1, explain why the circumference of a circle of radius r
on a sphere of radius R is $2\pi R \sin(r/R)$.

5.6 Negative Curvature

In Section 5.4 we defined the curvature of a circle of radius R to be $1/R$.
In 1665 Newton extended this idea to any smooth curve K by defining
the *curvature of K at a point P* to be the curvature of the circle that
"best approximates" K at P. (He took this to be the circle through P
whose center lies at the intersection of perpendiculars to the curve at
points infinitesimally close to P, and was able to find its curvature by
calculus.)

To define the curvature of a surface S at a point P we consider all
the planes perpendicular to S at P, and their curves of intersection,
called *sections* of S at P. Among these sections, there is one of maxi-
mum curvature κ_{max} and one of minimum curvature κ_{min}. These are
called the *principal curvatures*, and there are various ways in which
they may be combined to define a curvature of the surface. A good
combination is their product, $\kappa_{max}\kappa_{min}$, which is called the *Gaussian*
curvature of the surface S at point P. To explain why this is a good
concept of curvature we consider some examples.

If S is a sphere of radius R, then all sections of S, at any point, are circles of radius R. Thus the sphere has Gaussian curvature $1/R^2$ at all points, as we already mentioned in Section 5.4. There we also observed that the Gaussian curvature may be found from measurements *within* the sphere, and indeed this is the great virtue of Gaussian curvature: it is an *intrinsic* property of any smooth surface, a fact discovered by Gauss in 1827.

For a cylinder, the principal curvatures occur in the perpendicular sections shown in Figure 5.19. One of them is a straight line (a "circle of infinite radius"), which has curvature zero. Thus the Gaussian curvature of the cylinder is zero—like the plane—as it should be, since the cylinder is intrinsically a flat surface.

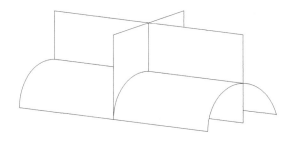

Figure 5.19: Principal curvature sections of the cylinder.

A sphere is a surface with constant positive Gaussian curvature, so a surface with properties "opposite" to those of the sphere should have constant *negative* curvature. Negative curvature makes sense, as a way to distinguish between surfaces like the sphere and surfaces like the *saddle* (Figure 5.20), which also has two nonzero principal curvatures.

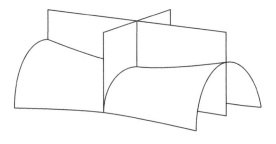

Figure 5.20: Principal curvature sections of a saddle.

On the sphere, the circles of principal curvature have their centers on the same side of the surface. On the saddle, the circles of principal curvature have their centers on opposite sides of the surface. We separate the two situations algebraically by giving the Gaussian curvature a negative sign when the two centers of curvature are on opposite sides. Thus the Gaussian curvature of a saddle is negative, and any negatively curved surface will look locally like a saddle.

Surfaces of constant negative curvature exist, but none of them is as easy to describe as the sphere. The best known example is called the *pseudosphere*, an infinite trumpet-shaped surface obtained by rotating a curve called the *tractrix* around the horizontal axis (Figure 5.21).

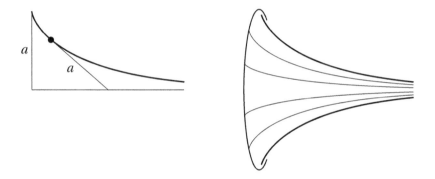

Figure 5.21: The tractrix and the pseudosphere.

The tractrix was first studied by Newton in 1676. It may be defined as a curve at constant tangential distance a from a line or, more mundanely, as the path of a stone pulled on a rope of length a by someone walking along a line (left picture in Figure 5.21). The tractrix has no smooth continuation past its endpoint on the left because its curvature becomes infinite there. The pseudosphere is the surface obtained by rotating the tractrix around the line, as shown in the right picture of Figure 5.21. It has a boundary circle traced by the endpoint of the rotating tractrix. Some nice properties of the tractrix and pseudosphere were discovered in the late seventeenth century, as mathematicians delighted in the new methods of calculus. But the real significance of the pseudosphere was recognized only in the 1830s, when its constant Gaussian curvature came to light.

In 1840, the German mathematician Ferdinand Minding worked out the relations between side lengths and angles of geodesic triangles on surfaces of constant Gaussian curvature, obtaining formulas *already known for triangles in the hypothetical non-Euclidean plane* (and in fact published by Lobachevsky, in the same journal, three years earlier)! Today, it is hard to comprehend why this did not cause a sensation, since it surely comes close to proving the reality of non-Euclidean geometry. Perhaps only Lobachevsky was interested, or perhaps Minding's result just wasn't close enough. It shows that the pseudosphere is *locally* like a non-Euclidean plane, and it also gives a realization of Saccheri's "asymptotic lines" as tractrix sections of the surface. But the pseudosphere is more like a non-Euclidean cylinder than a plane, and only half a cylinder at that, since it stops at the boundary circle.

5.7   The Hyperbolic Plane

Because of the shape of the pseudosphere, its "lines" extend to infinity in only one direction, which is a far cry from Euclid's requirement that all lines extend indefinitely. If one slits the pseudosphere along one of its tractrix sections the result is only an infinite wedge of the desired non-Euclidean plane. In fact, in ordinary space it is *impossible* to extend any constant negative curvature surface smoothly in all directions. This was not proved until 1901, but the obstacle was suspected in the nineteenth century, and in 1868 the Italian mathematician Eugenio Beltrami found an elegant way around it: instead of working with curved surfaces, he worked with images of them on the plane.

Beltrami started this train of thought in 1865, by asking which surfaces can be mapped onto the plane in such a way that their geodesics go to straight lines. He found that the answer is precisely the surfaces of constant Gaussian curvature. For example, great circles on the sphere can be mapped to lines on the plane, and the map that does the trick (to be precise, for the hemisphere) is *central projection,* shown in Figure 5.22. Rays from the center O to any great circle form a plane, which of course meets any other plane in a line. Thus projection from O to a plane sends great circles to lines, though only half of the sphere is mapped.

Figure 5.23 shows an example: a sphere divided into 48 spherical

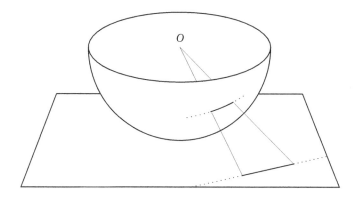

Figure 5.22: Central projection.

triangles and the image of its lower hemisphere under central projection, showing 24 regions with straight sides.

Figure 5.23: Triangulated sphere, centrally projected.

The situation with surfaces of negative curvature is just the opposite: all of the surface is mapped, but onto only part of the plane. In fact, the image always fits naturally into an open disk: a disk minus its boundary circle.

For example, the image of the pseudosphere is a wedge like that shown in Figure 5.24. The asymptotic tractrices on the pseudophere map to a wedge of line segments meeting at a point on the boundary of the disk, which can therefore be regarded as their common "point at infinity." The circular cross sections of the pseudosphere—which are *not* geodesics, but more like circles of latitude on a sphere—map to ellipses, tangential to the disk boundary at the endpoint of the wedge. (The dotted portions of the ellipses represent the result of "unrolling" the pseudosphere as one would unroll a cylinder.)

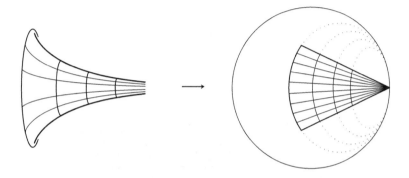

Figure 5.24: Geodesic-preserving map of the pseudosphere.

Clearly, the wedge does not fill the open disk, but it has a *natural extension* to the open disk. The pseudosphere can be "unrolled" all the way along the dotted ellipses (Beltrami did this by imagining a surface wrapped infinitely many times around the pseudosphere, and mapping the wrapping surface onto the disk), and each line segment can be extended backward to the other side of the disk.

The "distance" between points in the wedge is taken to be the distance between the corresponding points on the pseudosphere, and this notion of distance also extends naturally to the whole open disk. Let us call it *pseudodistance.* Since the line segments in the disk are images of geodesic segments on the pseudosphere, each line segment on the disk gives the shortest pseudodistance between its endpoints. Also, the pseudodistance from any point in the open disk to the boundary circle is infinite, because the length of the pseudosphere is infinite.

The open disk can thus be interpreted as an infinite "plane," whose "points" are the points of the disk, whose "lines" are line segments joining boundary points of the disk, and whose "distance" between "points" is pseudodistance. Each "line" is infinite and there is a unique "line" through any two "points." But the "plane" is non-Euclidean, as Figure 5.25 shows. For any "line" $\mathscr{L}$, and a "point" P outside it, there are many lines through P that do not meet $\mathscr{L}$.

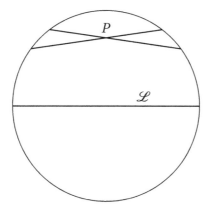

Figure 5.25: Parallels in the hyperbolic plane.

This non-Euclidean plane is called the *hyperbolic* plane. Or rather, it is a *model* of the hyperbolic plane, since there are various maps of the hyperbolic plane, like the various maps of the earth. But unlike the earth, which can be modelled by an actual globe, there is no physical hyperbolic plane. We know it only through its models, and there is no reason to choose a particular one as "real." The model just described is called the *projective model*, and indeed it resembles a projective view, inasmuch as "lines" look straight but distances and angles are distorted. There are also *conformal* views. They preserve angles and the shape of infinitesimal figures, but distort straightness of lines.

Figure 5.26 shows two views of the hyperbolic plane. On the top is *Circle Limit IV* by the Dutch artist M. C. Escher, which is a conformal view. You can see, for example, that wing tips meet everywhere at right angles.

On the bottom the picture is transformed to a projective view. In each picture the angels and devils are *all the same size* according to the hyperbolic notion of distance. We can estimate the hyperbolic length

Figure 5.26: The conformal and projective disk models.

of a given curve by counting the number of angels that lie along it, and the "line" between two points is a curve of minimal length; that is, the curve passing through the minimal number of angels and devils. In the top picture, these "lines" are circular arcs perpendicular to the boundary circle, and in the bottom picture they are ordinary line segments.

Notice that if we follow each "line" all the way to the boundary we will pass infinitely many angels and devils. Thus "lines" in the hyperbolic plane are infinite, as they are in the Euclidean plane.

Figure 5.27 shows another conformal view called the *half-plane model*. The "lines" in it are semicircles perpendicular to the boundary.

Figure 5.27: The half-plane model.

Exercises

Figure 5.23 shows a triangulated sphere centrally projected onto the plane, confirming that great circles are mapped to straight lines.

5.7.1 What are the angles in the straight triangles in the plane? Explain why this shows that central projection does *not* preserve angles.

This example raises an interesting question about triangles with angles π/l, π/m, π/n, for integers $l, m, n \geq 2$, which we call (l, m, n) triangles.

5.7.2 Show that if the (l, m, n) triangle lies on the Euclidean plane then

$$\frac{1}{l} + \frac{1}{m} + \frac{1}{n} = 1,$$

and find all solutions of this equation. (Hint: suppose $l \leq m \leq n$.)

5.7.3 Show that if the (l, m, n) triangle lies on the sphere then

$$\frac{1}{l} + \frac{1}{m} + \frac{1}{n} \geq 1,$$

and find all solutions of this inequality.

In the hyperbolic plane the area of a triangle, as on the sphere, is proportional to the difference between its angle sum and π. However, the angle sum of a hyperbolic triangle is *less* than π.

5.7.3 Deduce from this fact that the area of a hyperbolic triangle cannot be arbitrarily large.

5.7.5 Show that if an (l, m, n) triangle lies in the hyperbolic plane then

$$\frac{1}{l} + \frac{1}{m} + \frac{1}{n} \le 1.$$

5.7.6 Show that the triangle of smallest area among these is the $(2, 3, 7)$.

5.7.7 Verify, by counting numbers of corners at each vertex, that Figure 5.28 shows a tiling by $(2,3,7)$-triangles.

Figure 5.28: Tiling by $(2,3,7)$-triangles.

5.7.8 Show, by cutting a polygon into triangles, that the angle sum of a hyperbolic n-gon is less than the value $(n-2)\pi$ for a Euclidean n-gon, and that the area of the hyperbolic n-gon is proportional to the difference.

5.8 Hyperbolic Space

There is an analogous three-dimensional space of constant negative curvature, called *hyperbolic space*, and various models of it. The projective model is an open three-dimensional ball, with line segments joining points on the boundary sphere as "lines." A view showing what hyperbolic space looks like from the inside is shown in Figure 5.29.

Figure 5.29: Hyperbolic space.

The original version of this view, by Charlie Gunn, can be seen at

http://www.geom.uiuc.edu/graphics/pix/Video_Productions/
Not_Knot/NKposter.1500.html

I have made a negative of the original, because it seems clearer with a white sky rather than a black one. The "lines" in this view (the center lines of the beams) really look straight, but points *equidistant* from "lines" (on the edges of the beams) seem to lie on curves. This shows a curious property of the hyperbolic plane: the points at constant distance from a "line" do not lie on a "line." (A similar thing is true on the sphere: points equidistant from a great circle do not lie on a great circle; they lie on a smaller circle.) It is only in the Euclidean plane that the points equidistant from a line lie on a line.

The non-Euclidean nature of this space can also be seen from the angle sums of polygons. One can see many pentagons whose angles are all right angles, whereas a Euclidean regular pentagon has angles equal to $3\pi/5$.

5.9 Mathematical Space and Actual Space

> In the present work we have been interested mainly in offering a concrete counterpart of abstract geometry; however, we do not wish to omit a declaration that the validity of the new order of concepts does not depend on the possibility of such a counterpart.
>
> Eugenio Beltrami, Essay on the interpretation of non-Euclidean geometry, in Stillwell, *Sources of Hyperbolic Geometry*, p. 28.

The discovery of non-Euclidean geometry profoundly affected the development of mathematics in the nineteenth century, and also the development of physics in the twentieth century. It forced mathematicians to answer some questions they had previously been able to avoid:

- What is geometry?

- Is our mental image of space mathematically precise?

- Does actual space agree with our mental image?

- Is more than one geometry logically possible?

It is a little surprising that these questions did not affect mathematics earlier, for example, when the idea of a finite universe was in vogue in the middle ages. For whatever reason, the great advances in science and mathematics in the seventeenth and eighteenth centuries took place under a firm consensus that Euclid's geometry (*Euclidean* geometry as we now call it) is the geometry of actual space, and that no other space is possible. The great German philosopher Immanuel Kant (who was also a significant contributor to astronomy, being the originator, in 1755, of the "nebular hypothesis" on the genesis of the solar system) tried to explain this apparent agreement between geometry and reality by asserting that geometry is *synthetic a priori*. He believed he could show that the space of astronomy and mathematical space were both necessarily Euclidean.[1]

Kant's idea is a subtle one, and I'm not sure I can explain it properly. Luckily, I don't need to, because *the agreement it was supposed to explain does not exist.* We now know that the space of astronomy is neither spherical, Euclidean, or hyperbolic, but of *variable* curvature, though it takes extremely delicate measurements to establish this fact. The curvature of space is small where we live, but it becomes significant in regions of space now coming within the range of observation, such as black holes. Even in the neighborhood of the earth, space curvature is taken into account by certain precision instruments in common use, such as global positioning systems.

At any rate, long before the curvature of space was first detected, Beltrami's construction of the hyperbolic plane showed that more than one kind of geometry is possible. Beltrami *assumed* that Euclidean space exists, and constructed a non-Euclidean plane inside it, with nonstandard definitions of "line" and "distance" (namely, line segments in the unit disk and pseudodistance). This shows that the geometry of Bolyai and Lobachevsky is *logically as valid* as the geometry of Euclid: if there is a space in which "lines" and "distance" behave as Euclid thought they do, then there is also a surface in which "lines" and "distance" behave as Bolyai and Lobachevsky thought they might.

In particular, the parallel axiom is *not* a logical consequence of Eu-

---

[1]The story of Kant and non-Euclidean geometry has often been presented as a philosopher making a blunder and getting his nose rubbed in it by mathematicians. I like this kind of story as much as the next person, but I don't really believe that Kant is an instance of it. His alleged error was to give Euclidean geometry a privileged position, but mathematicians also give it a privileged position, as the geometry of a *flat* space. Non-Euclidean geometry is the geometry of curved surfaces or curved space, so one is perfectly entitled to say that its "lines" are not really "straight"—they just look straight in certain maps.

clid's other axioms, because the hyperbolic plane satisfies all of Euclid's axioms except the axiom of parallels. This *independence of the parallel axiom* was first stated clearly by the German mathematician Felix Klein, who in 1871 reconstructed Beltrami's projective model (and also spherical and Euclidean geometry) directly from projective geometry. In 1873 Klein [50, p. 111] described the situation as follows:

> ...non-Euclidean geometry is by no means intended to decide the validity of the parallel axiom, but only *whether the parallel axiom is a mathematical consequence of the remaining axioms of Euclid*; a question to which these investigations give a definite *no*. Because ... these remaining axioms suffice to construct a system of theories which includes Euclidean geometry merely as a special case.

I have not yet said exactly what Euclid's axioms were, because mathematicians did not give them much thought until Beltrami and Klein made their discoveries. As long as Euclidean space was thought to be the only space, "obvious" properties of it were taken for granted: sometimes stated as axioms, sometimes simply as unconscious assumptions. With the discovery of hyperbolic geometry, it became clear that there were at least two equally valid axiom systems: one with existence and uniqueness of parallels (Euclidean geometry), and one with existence and nonuniqueness of parallels (hyperbolic geometry). And conceivably, still more geometries had been overlooked, because many unstated assumptions were found in Euclid as soon as mathematicians looked at the *Elements* closely.

The situation cleared up very satisfactorily with the appearance of Hilbert's *Grundlagen der Geometrie* (foundations of geometry) in 1899. Hilbert produced a set of about 20 axioms, including the parallel axiom, from which all of Euclid's geometry follows. He also showed that replacing Euclid's parallel axiom (one parallel) by the Bolyai and Lobachevsky parallel axiom (more than one parallel) gives exactly the theorems of hyperbolic geometry. Thus *the parallel axiom is precisely what separates the geometry of zero curvature from the geometry of constant negative curvature.*

The Arithmetization of Geometry

As mentioned above, actual space is *not* of constant curvature. Its curvature varies in a way determined by the distribution of matter in it. The geometry of such a space is not easily captured by axioms describing the behavior of large scale objects such as "lines" (geodesics). However, a concise description is possible by means of equations relating infinitesimal line segments. The basic equations describe curvature in terms of infinitesimals, and they were given by Gauss's student Bernhard Riemann in 1854. Riemann's work was the inspiration for some of the greatest advances in geometry and physics in the past 150 years, from Beltrami's interpretation of non-Euclidean geometry to Albert Einstein's theory of gravity (general relativity), the theory that brought curved space into daily life with the global positioning system.

We saw some examples of infinitesimal geometry in Chapter 4, and I do not wish to push the idea further here. Suffice it to say that it depends on points having coordinates, and hence on *real numbers*, and indeed numbers are crucial for all mathematical physics.

From this perspective, the geometries of constant curvature are exceptionally simple in being approachable through axioms about "lines." But this does not mean that it is not enlightening to approach them through coordinates. The geometries of constant curvature (spherical, Euclidean, hyperbolic) also stand out as coordinate geometries because of their *algebraic* character. As we saw in Section 4.5, the coordinates of points on a Euclidean straight line satisfy a so-called *linear* equation. The theory of linear equations, called *linear algebra*, embraces all of Euclidean geometry. Indeed, it embraces projective geometry as well, and hence also spherical and hyperbolic geometry through their projective models.

When real numbers are used as coordinates, the number of coordinates is the *dimension* of the geometry. This is why we call the plane two-dimensional and space three-dimensional. However, one can also expect complex numbers to be useful, knowing their geometric properties from Chapter 2. What is remarkable is that complex numbers are if anything *more* appropriate for spherical and hyperbolic geometry than for Euclidean geometry. With hindsight, it is even possible to see hyperbolic geometry in properties of complex numbers that were studied as early as 1800, long before hyperbolic geometry was discussed by anyone. This was noticed by the third great contributor to non-Euclidean geometry after Beltrami and Klein—the French mathematician Henri Poincaré (1854–1912).

It would take us too far afield to explain Poincaré's contributions here, and we refer the interested reader to *Sources of Hyperbolic Geometry*. However, we pursue the geometric content of complex numbers further in the next chapter, by taking up the question: are there three-dimensional numbers?

Chapter 6

The Fourth Dimension

Preview

The idea that space is flat, tenacious though it is, is more easily over-
come than the idea that space is three-dimensional. We can visualize
curved space, but *it seems impossible to visualize a direction perpendicular to three directions already perpendicular to each other.* Perhaps
because of this, four-dimensional geometry came to light later than
non-Euclidean geometry—in the 1840s.

Even then, the fourth dimension arose quite accidentally, from a
failed attempt to create *three-dimensional numbers.* Two-dimensional
numbers were known (the complex numbers), and they were known
to have the *multiplicative property* of absolute value: $|u||v| = |uv|$. *But
in more than two dimensions, a multiplicative absolute value implies
the seeming impossibility described above: four mutually perpendicular directions. In three dimensions this* is *impossible! But there is a con-
solation prize.* It becomes clear that one should not add and multiply
triples, but *quadruples.*

The result is the four-dimensional arithmetic of *quaternions.* This
system has most, though not all, the properties of real and complex
numbers. It is called four-dimensional simply because quaternions
have four coordinates, and one need not try to visualize the four perpendicular coordinate axes. Still, there is an irresistible urge to use geometric language when discussing this four-dimensional space.

In the first place, quaternions give a nice approach to symmetric

173

objects in three-dimensional space: the *regular polyhedra*. But this leads in turn to the *regular polytopes*, a family of four-dimensional symmetrical objects as remarkable as the regular polyhedra. *One then becomes convinced that four-dimensional space is not just a set of quadruples; it is a world of genuine geometry.*

6.1   Arithmetic of Pairs

In Chapter 2 we briefly mentioned that complex numbers $a + bi$ can be regarded as *ordered pairs of real numbers* (a, b), before moving on to the interpretation of (a, b) as a point in the plane, and its geometric implications. It is worth dwelling a little longer on the idea of adding and multiplying pairs of real numbers, because the ability to do these operations raises the possibility of doing them for triples, quadruples, and so on.

It was the Irish mathematician William Rowan Hamilton who in 1835 first suggested treating complex numbers as pairs of real numbers. The idea has merit in reducing properties of complex numbers to properties of reals—and thus avoiding the mysterious $\sqrt{-1}$—but it does not tell or suggest anything new about complex numbers. Indeed, Hamilton's agenda was actually to find an arithmetic of *triples* (and after that, quadruples, quintuples, and so on). He hoped that his arithmetic of pairs would suggest a general rule for doing arithmetic on n-tuples for any positive integer n.

Hamilton's arithmetic of pairs has addition defined by

$$(a_1, b_1) + (a_2, b_2) = (a_1 + a_2, b_1 + b_2)$$

and multiplication defined by

$$(a_1, b_1)(a_2, b_2) = (a_1 a_2 - b_1 b_2, a_1 b_2 + b_1 a_2),$$

as we saw in Section 2.5. The rules for adding and multiplying pairs are simply the rules for adding and multiplying complex numbers, with each complex number $a + bi$ rewritten as (a, b). They therefore satisfy the rules of algebra from Section 3.7, because the complex numbers do. For the same reason, the absolute value of a pair, $|(a, b)| = \sqrt{a^2 + b^2}$, has the *multiplicative property*

$$|(a_1, b_1)||(a_2, b_2)| = |(a_1 a_2 - b_1 b_2, a_1 b_2 + b_1 a_2)|,$$

and this is equivalent to the *two-square identity* of Diophantus:

$$(a_1^2 + b_1^2)(a_2^2 + b_2^2) = (a_1 a_2 - b_1 b_2)^2 + (a_1 b_2 + b_1 a_2)^2.$$

There is a natural way to add triples so that the laws of algebra for addition are satisfied, as we shall see in the next section. Thus the main problem in the algebra of triples is to *define a multiplication satisfying the laws of algebra and so that the absolute value is multiplicative.* Hamilton searched for a definition for at least 13 years, but what he wanted was impossible!

The situation is more difficult than he ever knew, because a multiplication with these properties is also lacking for quadruples, quintuples, and in fact all n-tuples for higher values of n. Despite this, Hamilton salvaged something of value—an arithmetic of quadruples called the *quaternions*, which satisfies all but one of the laws of algebra and has a multiplicative absolute value.

Hamilton's discovery of the quaternions can be appreciated better today than in his own time because we now see that they are "nearly impossible." The quaternions are an amazing rarity (and from the n-dimensional point of view, so are the real and complex numbers). We can show that there is nothing like them for higher values of n, and only for $n = 8$ does anything come close (see Section 6.5). In particular, it is now quite easy to see why there is no reasonable algebra of triples.

Reasons for this three-dimensional impossibility will be discussed in the next two sections.

Exercises

The formula

$$(a_1^2 + b_1^2)(a_2^2 + b_2^2) = (a_1 a_2 - b_1 b_2)^2 + (a_1 b_2 + a_2 b_1)^2 \qquad (*)$$

gives a way to write a large number as sum of two squares when it can be factorized into numbers known to be sums of two squares.

For example, we know that $13 = 3^2 + 2^2$ and therefore

$$169 = 13 \times 13 = (3^2 + 2^2)(3^2 + 2^2).$$

If we take $a_1 = 3, b_1 = 2, a_2 = 3, b_2 = 2$, the formula (*) allows us to write

$$(3^2 + 2^2)(3^2 + 2^2) = (3 \times 3 - 2 \times 2)^2 + (3 \times 2 + 3 \times 2)^2,$$

that is,

$$169 = 5^2 + 12^2.$$

6.1.1 Given what we have just found, namely $13^2 = 5^2 + 12^2$, write the number $13^4 = (5^2 + 12^2)(5^2 + 12^2)$ as a sum of two squares.

6.1.2 Using your answer to Question 6.1.1, write 13^8 as a sum of two squares.

6.1.3 Write 101×145 as a sum of two squares.

6.2 Searching for an Arithmetic of Triples

> Every morning in the early part of October 1843, on my coming down to breakfast, your brother William Edward and yourself used to ask me: "Well, Papa, can you multiply triples?" Whereto I was always obliged to reply, with a sad shake of the head, "No, I can only add and subtract them."
>
> Hamilton, letter to his son Archibald, 5 August 1865, in R. P. Graves, *The Life of Sir William Rowan Hamilton*, vol. II, p. 435.

Hamilton realized that there is a simple and natural way to add triples, namely *vector addition*, extending addition of single real numbers and pairs: if $v_1 = (a_1, b_1, c_1)$ and $v_2 = (a_2, b_2, c_2)$ then

$$v_1 + v_2 = (a_1 + a_2, b_1 + b_2, c_1 + c_2).$$

This definition of addition obviously generalizes to n-tuples and has the same algebraic properties as addition of numbers:

$$u + v = v + u$$
$$u + (v + w) = (u + v) + w$$
$$u + \mathbf{0} = u$$
$$u + (-u) = \mathbf{0}.$$

Compare these equations with the laws of algebra listed in Section 3.7. Here, $\mathbf{0}$ represents the *zero vector* $(0,0,0)$ and $-u = (-a, -b, -c)$ is the *additive inverse* or *opposite* of the vector $u = (a, b, c)$.

The hard part is to define a multiplication that satisfies the laws of algebra and so that the absolute value $|v| = \sqrt{a^2 + b^2 + c^2}$ of $v = (a, b, c)$ has the multiplicative property:

$$|v_1|^2 |v_2|^2 = |v_1 v_2|^2,\tag{1}$$

where

$$v_1 = (a_1, b_1, c_1) \quad \text{and hence} \quad |v_1|^2 = a_1^2 + b_1^2 + c_1^2$$

and

$$v_2 = (a_2, b_2, c_2) \quad \text{and hence} \quad |v_2|^2 = a_2^2 + b_2^2 + c_2^2.$$

Thus if $v_1 v_2 = (a, b, c)$, the multiplicative property (1) gives

$$(a_1^2 + b_1^2 + c_1^2)(a_2^2 + b_2^2 + c_2^2) = a^2 + b^2 + c^2,\tag{2}$$

where a, b, and c are some functions of $a_1, b_1, c_1, a_2, b_2, c_2$. It is uncertain what these functions may be, but one thing is clear immediately: *if a, b, and c are integers whenever $a_1, b_1, c_1, a_2, b_2, c_2$ are integers, then such functions do not exist.* If they exist, then there must be a *three-square identity*, and in particular

$$(1^2 + 1^2 + 1^2)(0^2 + 1^2 + 2^2) = 15 = a^2 + b^2 + c^2 \quad \text{for some integers } a, b, \text{ and } c.$$

But this is false, as can be seen by checking all the sums $a^2 + b^2 + c^2$ for positive integers $a, b, c < 15$.

This example rules out any simple product of triples analogous to Hamilton's product of pairs. Hamilton was not aware of it, because he searched for integer-valued products without success for many years. Evidently he had not read enough of the literature in number theory, because sums of three squares had been studied decades earlier, by the French mathematician Adrien-Marie Legendre, who pointed out in his *Théorie des Nombres* that:

$$(1^2 + 1^2 + 1^2)(1^2 + 2^2 + 4^2) = 63 \neq a^2 + b^2 + c^2 \quad \text{for integers } a, b, \text{ and } c.$$

(This example is the smallest one for which the factors are sums of three *nonzero* squares.)

Thus simple arithmetic makes a product of triples with multiplicative absolute value look unlikely, even if it doesn't rule it out completely. In the next section we give a conclusive proof, which rules out the possibility of such a product of n-tuples for any $n > 2$.

Exercises

The numbers 15 and 63 are both of the form $8n+7$, and in fact *no* numbers of this form are sums of three squares. However, many of them are products of sums of three squares.

6.2.1 Factorize 39 into two numbers and write the factors as sums of three squares (you may include 0^2 among the squares).

6.2.2 Similarly factorize 55, 87, and 95.

It is possible to explain why each number $8n+7$ is not a sum of three squares by considering remainders on division by 8.

6.2.3 Explain why any natural number is of one of the forms $8n$, $8n\pm1$, $8n\pm2$, $8n\pm3$, $8n+4$, where n is a natural number.

6.2.4 Show that the square of a number of the form $8n$, $8n\pm1$, $8n\pm2$, $8n\pm3$, $8n+4$ leaves remainder $0, 1, 4, 1, 0$, respectively, on division by 8.

6.2.5 Deduce from Question 6.2.4 that a sum of three squares leaves remainder 0, 1, 2, 3, 4, 5, or 6 on division by 8. (Hence a number of the form $8n+7$ is *not* a sum of three squares.)

6.3 Why n-tuples Are Unlike Numbers when $n \geq 3$

To refute products of triples conclusively we consider *geometric* consequences of the multiplicative absolute value, as we did for the product of pairs in Section 2.6. We view the numbers a, b, c in the triple (a, b, c) as coordinates of a point located in Euclidean three-dimensional space as shown in Figure 6.1.

Then the absolute value $|(a, b, c)| = \sqrt{a^2 + b^2 + c^2}$ is the distance of the point $P = (a, b, c)$ from the origin O. This follows from Figure 6.1 by the Pythagorean theorem: OQ is the hypotenuse of the right-angled triangle ORQ with sides a and b, hence $OQ = \sqrt{a^2 + b^2}$. And OP is the hypotenuse of the right-angled triangle OQP with sides $\sqrt{a^2 + b^2}$ and c, hence $OP = \sqrt{a^2 + b^2 + c^2}$.

It follows, as for pairs in Section 2.6, that for any triples v and w,

$$|w - v| = \text{distance between } v \text{ and } w.$$

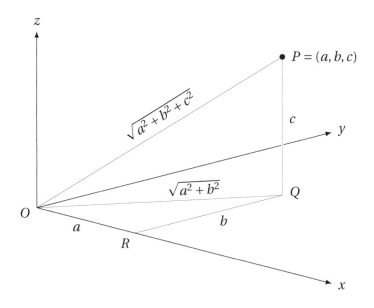

Figure 6.1: Coordinates and distance in three dimensions.

And if the product of triples satisfies the distributive law,

$$u(w - v) = uw - uv,$$

then the multiplicative property of absolute value implies

distance between uv and $uw = |uw - uv| = |u(w - v)| = |u||w - v|$

$$= |u| \times \text{distance between } v \text{ and } w.$$

Thus *if all points in three-dimensional space are multiplied by the point u then all distances are multiplied by the constant number* $|u|$. When $|u| = 1$, this means all distances are unchanged. That is, *when multiplied by u with* $|u| = 1$, *space is moved rigidly*. In particular, angles are preserved.

Now suppose that there is a product of triples which, together with vector addition, satisfies all the laws of algebra. In particular, there is a *multiplicative identity*: a point **1** such that $|1| = 1$ and $u1 = u$ for any triple u. Moreover, since space is three-dimensional, we can find points **i** and **j**, also of absolute value 1, such that **1**, **i**, and **j** are in mutually perpendicular directions from O. Figure 6.2 shows these points, together with their negatives.

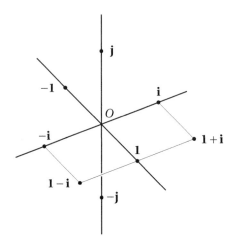

Figure 6.2: Points in perpendicular directions from O.

Figure 6.2 also shows the point $\mathbf{1}+\mathbf{i}$, whose distance from O is clearly $\sqrt{2}$, the diagonal of a unit square. Thus $|\mathbf{1}+\mathbf{i}| = \sqrt{2}$, and similarly $|\mathbf{1}-\mathbf{i}| = \sqrt{2}$. It then follows from the multiplicative property of absolute value, and the laws of algebra, that

$$2 = |\mathbf{1}+\mathbf{i}||\mathbf{1}-\mathbf{i}| = |(\mathbf{1}+\mathbf{i})(\mathbf{1}-\mathbf{i})| = |\mathbf{1}-\mathbf{i}^2|.$$

This says that the point $\mathbf{1}-\mathbf{i}^2$ is at distance 2 from O. We also know, by the multiplicative property of absolute value again, that $|\mathbf{i}^2| = |\mathbf{i}|^2 = 1^2 = 1$. Thus the point $\mathbf{1}-\mathbf{i}^2$ is at distance 1 from $\mathbf{1}$. But the *only* point at distance 2 from O and distance 1 from $\mathbf{1}$ is $\mathbf{1}+\mathbf{1}$, so we must have $\mathbf{i}^2 = -\mathbf{1}$.

A similar argument shows that $\mathbf{j}^2 = -\mathbf{1}$, and more generally that $u^2 = -\mathbf{1}$ for any point u whose direction from O is perpendicular to the direction of $\mathbf{1}$, and whose absolute value is 1.

This is suspicious. In ordinary algebra, whoever heard of so many square roots of -1? Indeed, by the factor theorem mentioned in Section 2.7, square roots of -1 correspond to factors of $x^2 + 1$, and there can be at most two of these. A contradiction is looming, and we run smack into it by taking the product $\mathbf{ij}$. We do not know exactly what $\mathbf{ij}$ is, but we can tell that $\mathbf{ij}$ and $\mathbf{1}$ are in perpendicular directions from O.

Why? Because **i** and **j** are in perpendicular directions, hence (multiplying the whole space by **i**) so are $\mathbf{i}^2 = -1$ and **ij**.

But if -1 and **ij** are in perpendicular directions, so are **1** and **ij**.

Thus **ij** is one of the points u for which $u^2 = -1$. Let us see where this leads, assuming the laws of algebra:

$$-1 = (\mathbf{ij})^2 = (\mathbf{ij})(\mathbf{ij}) = \mathbf{jiij} \qquad \text{by commutative and associative laws}$$
$$= \mathbf{j}(-1)\mathbf{j} \qquad \text{since } \mathbf{i}^2 = -1$$
$$= -\mathbf{j}^2 \qquad \text{by commutative and associative laws}$$
$$= 1 \qquad \text{since } \mathbf{j}^2 = -1.$$

Contradiction! Therefore there is no product of triples satisfying all the laws of algebra.   □

In the argument above we used a picture of three dimensions, but all we actually *assumed* about space was

- distance given by absolute value,

- existence of at least three mutually perpendicular directions, and

- an instance of the so-called *triangle inequality*: that travelling distance 1 in one direction, then distance 1 in a different direction, results in total travel through distance less than 2.

These properties hold in *Euclidean space* $\mathbb{R}^n$ of dimension $n \geq 3$. This space is defined to be the set of n-tuples $(x_1, \ldots, x_n)$ of real numbers, with the vector addition operation and distance between points $(x_1, \ldots, x_n)$ and $(x'_1, \ldots, x'_n)$ given by

$$\left| (x'_1, \ldots, x'_n) - (x_1, \ldots, x_n) \right| = \sqrt{(x'_1 - x_1)^2 + \cdots + (x'_n - x_n)^2}.$$

Thus our argument also shows that *there is no product operation for points of $\mathbb{R}^n$, when $n \geq 3$, for which the absolute value is multiplicative and all the laws of algebra hold.*

To put it another way, ordinary algebra is *impossible* in any Euclidean space of three or more dimensions. But from the ashes of ordinary algebra there arises an extraordinary algebra in four dimensions.

Exercises

Given that $u^2 = -1$ for any unit vector u in the plane of $\mathbf{i}$ and $\mathbf{j}$, another way to get a contradiction (to the assumption of 3-dimensional numbers) is the following.

6.3.1 Explain geometrically why $\left|\frac{1}{\sqrt{2}}(\mathbf{i}+\mathbf{j})\right| = 1$. (So $\frac{1}{\sqrt{2}}(\mathbf{i}+\mathbf{j})$ is unit vector in the plane of $\mathbf{i}$ and $\mathbf{j}$, and hence its square equals -1.)

6.3.2 Assuming that all the laws of algebra hold, deduce from Exercise 6.3.1 that
$$-1 = \frac{1}{2}(\mathbf{i}^2 + \mathbf{j}^2 + 2\mathbf{ij}),$$
and hence that $\mathbf{ij} = 0$.

6.3.3 This contradicts the multiplicative property $|uv| = |u||v|$. Why?

6.4 Quaternions

The contradiction found above,
$$-1 = (\mathbf{ij})(\mathbf{ij}) = \mathbf{ijji} = \mathbf{i}(-1)\mathbf{i} = -\mathbf{ii} = 1,$$

can be averted only if we sacrifice at least one of the laws of algebra. Hamilton arrived at a similar impasse in 1843, and decided that the least damaging way around the obstacle was to abandon commutative multiplication. This is because the contradiction is averted by assuming
$$\mathbf{ij} = -\mathbf{ji},$$

and no other contradictions arise after this assumption is made. However, it is still a mystery what the product $\mathbf{ij}$ represents.

After a series of computational experiments, motivated partly by algebra and partly by geometry, Hamilton became convinced that $\mathbf{ij}$ lies in the *fourth dimension*, perpendicular to the directions of $\mathbf{1}, \mathbf{i}$, and $\mathbf{j}$. He believed this without knowing the most important consequence of the multiplicative absolute value, namely, that multiplication of the whole space by a point of absolute value 1 is a rigid motion. Once this is recognized (as we did in the previous section), Hamilton's tortuous path of discovery can be straightened out as follows.

Assume, as in Section 6.3, that **1** is the multiplicative identity and choose points **i** and **j** in perpendicular directions from O that are also perpendicular to the direction of **1**. We already know that the direction of **ij** is perpendicular to the direction of **1**. It is even easier to show that **ij** is perpendicular to the directions of **i** and **j**. We consider two rigid motions of the whole space, induced by suitable multiplications:

- The first motion sends the pair **1**, **i** to the pair **j**, **ij** by moving each point u to the point $u\mathbf{j}$. Since the directions of **1** and **i** are perpendicular, so are the directions of **j** and **ij**.

- The second motion sends the pair **1**, **j** to the pair **i**, **ij** by moving each point u to the point $i u$. Since the directions of **1** and **j** are perpendicular, so are the directions of **i** and **ij**.

Thus **ij**, which Hamilton called **k**, lies in a "fourth dimension." Its direction is perpendicular to all directions in the three-dimensional space spanned by **1**, **i**, and **j**. Hamilton realized that accepting **k** is a bold step from the geometric point of view, but algebraically it amounts to no more than considering *quadruples* of real numbers. Any combination $a\mathbf{1} + b\mathbf{i} + c\mathbf{j} + d\mathbf{k}$ of the *basis elements* **1**, **i**, **j**, **k** can be identified with the quadruple (a, b, c, d), just as a complex number $a + bi$ can be identified with the pair (a, b).

Addition of quadruples is the usual *vector addition*,

$$(a_1, b_1, c_1, d_1) + (a_2, b_2, c_2, d_2) = (a_1 + a_2, b_1 + b_2, c_1 + c_2, d_1 + d_2),$$

because this reflects the result of adding combinations of basis elements according to the laws of addition and the distributive law:

$$(a_1\mathbf{1} + b_1\mathbf{i} + c_1\mathbf{j} + d_1\mathbf{k}) + (a_2\mathbf{1} + b_2\mathbf{i} + c_2\mathbf{j} + d_2\mathbf{k})$$
$$= a_1\mathbf{1} + a_2\mathbf{1} + b_1\mathbf{i} + b_2\mathbf{i} + c_2\mathbf{j} + c_2\mathbf{j} + d_1\mathbf{k} + d_2\mathbf{k}$$
$$= (a_1 + a_2)\mathbf{1} + (b_1 + b_2)\mathbf{i} + (c_1 + c_2)\mathbf{j} + (d_1 + d_2)\mathbf{k}.$$

But what is $a_1\mathbf{1} + b_1\mathbf{i} + c_1\mathbf{j} + d_1\mathbf{k}$ times $a_2\mathbf{1} + b_2\mathbf{i} + c_2\mathbf{j} + d_2\mathbf{k}$? It is not even clear that the product lies in the space of quadruples—what if **jk**, say, lies in a fifth direction, perpendicular to the directions of **1**, **i**, **j**, and **k**? Fortunately, this does not happen. The other products follow from the known ones by the laws of algebra, without using commutativity of multiplication, and they are all combinations of basis elements.

For example, the two products equal to $\mathbf{k}$, $\mathbf{ij}$ and $-\mathbf{ji}$, give us the value of $\mathbf{jk}$ as follows:

$$
\begin{aligned}
\mathbf{jk} &= \mathbf{j(ij)} && \text{since } \mathbf{k} = \mathbf{ij} \\
&= \mathbf{j(-ji)} && \text{since } \mathbf{ij} = -\mathbf{ji} \\
&= -(\mathbf{jj})\mathbf{i} && \text{by associativity} \\
&= \mathbf{i} && \text{since } \mathbf{j}^2 = -1.
\end{aligned}
$$

We similarly find that $\mathbf{kj} = -\mathbf{i}$ and that $\mathbf{ki} = \mathbf{j} = -\mathbf{ik}$.

The value of any product $(a_1\mathbf{1} + b_1\mathbf{i} + c_1\mathbf{j} + d_1\mathbf{k})(a_2\mathbf{1} + b_2\mathbf{i} + c_2\mathbf{j} + d_2\mathbf{k})$ now follows by the distributive law. (Or more precisely, by left and right distributive laws. We need both when multiplication is not commutative.) The computation is long but straightforward:

$$
\begin{aligned}
(a_1\mathbf{1} + b_1\mathbf{i} + c_1\mathbf{j} + d_1\mathbf{k})&(a_2\mathbf{1} + b_2\mathbf{i}+c_2\mathbf{j} + d_2\mathbf{k}) \\
= \quad & (a_1\mathbf{1} + b_1\mathbf{i} + c_1\mathbf{j} + d_1\mathbf{k})a_2\mathbf{1} \\
+ \, & (a_1\mathbf{1} + b_1\mathbf{i} + c_1\mathbf{j} + d_1\mathbf{k})b_2\mathbf{i} \\
+ \, & (a_1\mathbf{1} + b_1\mathbf{i} + c_1\mathbf{j} + d_1\mathbf{k})c_2\mathbf{j} \\
+ \, & (a_1\mathbf{1} + b_1\mathbf{i} + c_1\mathbf{j} + d_1\mathbf{k})d_2\mathbf{k}
\end{aligned}
$$

using the left distributive law $u(v + w) = uv + uw$

$$
\begin{aligned}
= \quad & (a_1 a_2 - b_1 b_2 - c_1 c_2 - d_1 d_2)\mathbf{1} \\
+ \, & (a_1 b_2 + b_1 a_2 + c_1 d_2 - d_1 c_2)\mathbf{i} \\
+ \, & (a_1 c_2 - b_1 d_2 + c_1 a_2 + d_1 b_2)\mathbf{j} \\
+ \, & (a_1 d_2 + b_1 c_2 - c_1 b_2 + d_1 a_2)\mathbf{k}
\end{aligned}
$$

using the right distributive law $(u + v)w = uw + vw$.

One need not remember this complicated rule, because it follows from the rules for multiplying the basis elements $\mathbf{1}$, $\mathbf{i}$, $\mathbf{j}$, $\mathbf{k}$. The latter rules can be boiled down to the following equations, discovered by Hamilton on 16 October 1843:

$$
\mathbf{i}^2 = \mathbf{j}^2 = \mathbf{k}^2 = \mathbf{ijk} = -1.
$$

With these equations, *Hamilton had defined a multiplication operation on* $\mathbb{R}^4$ *that satisfies all the laws of algebra except the commutative law* $uv = vu$.

He was so elated by this discovery that he carved the equations
on a bridge he happened to be passing when the idea came to him:
Broombridge in Dublin. The carving disappeared long ago, but the
bridge now carries a plaque commemorating the event. See Figure 6.3,
and also Robert Burke's web site "A Quaternion Day Out" (the URL has
changed, so it is probably best to find it with a search engine) for this
and other nice photographs of the location.

Figure 6.3: The quaternion plaque on Broombridge.

The plaque itself is somewhat worn, but this is what it says:

> Here as he walked by
> on the 16th of October 1843
> Sir William Rowan Hamilton
> in a flash of genius discovered
> the fundamental formula for
> quaternion multiplication
> $$i^2 = j^2 = k^2 = ijk = -1$$
> & cut it on a stone of this bridge

Hamilton called the system of quadruples, with their rules for addition and multiplication, the *quaternions*. Elated as he was by the success of his decision to let $\mathbf{ij} = -\mathbf{ji}$—it made up for the long years of failure trying to multiply triples—his mind was still troubled by one question: does the absolute value have the multiplicative property? After all, this was his original demand, so the success of his whole speculation hung on a positive answer.

Exercises

6.4.1 Show that each of the primes 2, 3, 5, 7, 11, 13, 17, 19, 23, 29, and 31 is a sum of four squares (some of which may be zero).

6.4.2 Which of these primes are not sums of three squares?

The rules for multiplying different quaternions among **i**, **j**, and **k** are summarized by Figure 6.4. The product of any two distinct quater-

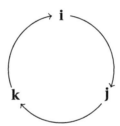

Figure 6.4: Products of the imaginary quaternion units.

nions is the third quaternion in the circle, with a + sign if an arrow points from the first element to the second, and a − sign otherwise. For example $\mathbf{ij} = \mathbf{k}$, but $\mathbf{ji} = -\mathbf{k}$.

The rule for multiplying any of **i**, **j**, **k** by *itself* is $\mathbf{i}^2 = \mathbf{j}^2 = \mathbf{k}^2 = -1$.

6.4.3 Check that $\mathbf{i}(\mathbf{jk}) = (\mathbf{ij})\mathbf{k} = -1$.

6.4.4 $(\mathbf{i}+\mathbf{j})(\mathbf{i}-\mathbf{j})$ is *not* equal to $\mathbf{i}^2 - \mathbf{j}^2$. What is the correct value?

6.4.5 Show that $(\mathbf{i}+\mathbf{j})^2 = -2$.

6.5   The Four-Square Theorem

Hamilton made the leap into the fourth dimension to escape contradictions in three-dimensional algebra, such as the existence of four
points 1, $\mathbf{i}$, $\mathbf{j}$, and $\mathbf{ij}$ in mutually perpendicular directions. He was di-
rected there by *consequences* of a multiplicative absolute value, but
without knowing that such an absolute value actually exists. To complete his escape he needed to define quaternion absolute value and to
show that it has the multiplicative property $|u||v| = |uv|$ for any quater-
nions u and v.

It is not hard to determine what $|a\mathbf{1} + b\mathbf{i} + c\mathbf{j} + d\mathbf{k}|$ should be. In
$\mathbb{R}^4$, the point $a\mathbf{1} + b\mathbf{i} + c\mathbf{j} + d\mathbf{k}$ is found at distance d in the direction $\mathbf{k}$,
which is perpendicular to $a\mathbf{1} + b\mathbf{i} + c\mathbf{j}$. And $a\mathbf{1} + b\mathbf{i} + c\mathbf{j}$ lies at distance
$\sqrt{a^2 + b^2 + c^2}$ from O, as we saw in Section 6.3. This gives a right-angled
triangle with sides $\sqrt{a^2 + b^2 + c^2}$ and d, so the Pythagorean theorem
implies that $a\mathbf{1} + b\mathbf{i} + c\mathbf{j} + d\mathbf{k}$ is at distance $\sqrt{a^2 + b^2 + c^2 + d^2}$ from O.
Thus the appropriate definition of quaternion absolute value is

$$|a\mathbf{1} + b\mathbf{i} + c\mathbf{j} + d\mathbf{k}| = \sqrt{a^2 + b^2 + c^2 + d^2}.$$

To avoid square roots, we can write the multiplicative property as
$|u|^2|v|^2 = |uv|^2$. Then if we put

$$u = a_1\mathbf{1} + b_1\mathbf{i} + c_1\mathbf{j} + d_1\mathbf{k},$$
$$v = a_2\mathbf{1} + b_2\mathbf{i} + c_2\mathbf{j} + d_2\mathbf{k},$$

and take uv from the quaternion product formula of the last section,
the multiplicative property becomes the following *four-square identity*:

$$
\begin{aligned}
(a_1^2 + b_1^2 + c_1^2 + d_1^2)(a_2^2 + b_2^2 + c_2^2 + d_2^2) = {} & (a_1 a_2 - b_1 b_2 - c_1 c_2 - d_1 d_2)^2 \\
& + (a_1 b_2 + b_1 a_2 + c_1 d_2 - d_1 c_2)^2 \\
& + (a_1 c_2 - b_1 d_2 + c_1 a_2 + d_1 b_2)^2 \\
& + (a_1 d_2 + b_1 c_2 - c_1 b_2 + d_1 a_2)^2.
\end{aligned}
$$

Hamilton had never heard of such a formula, and it was with some
amazement that he expanded the right side and watched 24 terms cancel, leaving just the 16 terms in the expanded left side. Thus the quaternion absolute value *is* multiplicative, and it is relatively straightforward

to verify all the other laws of algebra, except of course the commutative law of multiplication.

Hamilton's dream of three-dimensional numbers was indeed impossible, but the reality turned out to be more interesting. The known systems of numbers (real and complex) are *exceptional structures*, existing only in one and two dimensions, and the system of quaternions is even more exceptional. It is the only n-dimensional structure satisfying all the laws of algebra except commutative multiplication. This was first proved by the German mathematician Georg Frobenius in 1878, but unfortunately Hamilton did not live long enough to see it.

He did live long enough, however, to fill the gaps in his knowledge about sums of three and four squares. The day after he discovered the quaternions and the four square identity he described them to his friend John Graves, with whom he often discussed the algebra of n-tuples. Within a couple of months Graves found a similar *eight-square identity*, and with it an algebra of octuples, now called the *octonions*.

The octonions satisfy all the laws of algebra except $uv = vu$ and $u(vw) = (uv)w$, so they are an interesting story in themselves—though unfortunately too long to tell here. In the history of n-dimensional algebra they are noteworthy for showing, finally, that *sums of squares are the key issue*.

Graves evidently saw that the four-square identity encapsulates the whole algebra of quaternions, just as Diophantus' two-square identity encapsulates the algebra of complex numbers. Also, one should never have expected an algebra of triples because there is no three-square identity. Graves consulted the number theory literature and reported back to Hamilton as follows:

> On Friday last I looked into Lagrange's [he meant Legendre] Théorie des Nombres and found for the first time that I had lately been on the track of former mathematicians. For example, the mode by which I satisfied myself that a general theorem
>
> $$(x_1^2 + x_2^2 + x_3^2)(y_1^2 + y_2^2 + y_3^2) = z_1^2 + z_2^2 + z_3^2$$
>
> was impossible was the very mode mentioned by Legendre, who gives the very example that occurred to me, viz.,

$3 \times 21 = 63$, it being impossible to compound 63 of three squares.

I then learned that the theorem

$$(x_1^2 + x_2^2 + x_3^2 + x_4^2)(y_1^2 + y_2^2 + y_3^2 + y_4^2) = z_1^2 + z_2^2 + z_3^2 + z_4^2$$

was Euler's.

Graves, letter to Hamilton 1844, in Hamilton's *Mathematical Papers*, page 649. (The example is on page 184 of Legendre's 1808 edition.)

Where Were Quaternions before 1843?

Euler discovered the four-square identity in 1748, so this could be considered the first "sighting" of quaternions, analogous to the "sightings" of complex numbers in the two-square identity. Indeed, the parallel between complex numbers and quaternions goes further than this. Like the complex numbers, the quaternions give a natural representation of rotations, and certain representations of space rotations, discovered before 1843, looked a lot like quaternions after 1843. The first "quaternion-like" representation of rotations was found by Gauss, in about 1819, and another was found in 1840 by the French mathematician Olinde Rodrigues.

Gauss also observed "quaternion" behavior in *pairs of complex numbers*, analogous to the "complex" behavior of pairs of real numbers in the Diophantus identity. He found that the four-square identity is equivalent to the following *complex two-square identity*:

$$\left(|u_1|^2 + |v_1|^2\right)\left(|u_2|^2 + |v_2|^2\right) = \left|u_1 u_2 - v_1 \overline{v_2}\right|^2 + \left|u_1 v_2 + v_1 \overline{u_2}\right|^2,$$

where the bar denotes the *complex conjugate* defined by $\overline{a + bi} = a - bi$. We now know that quaternions can indeed be defined as pairs (u, v) of complex numbers, with the multiplication rule

$$(u_1, v_1)(u_2, v_2) = (u_1 u_2 - v_1 \overline{v_2}, u_1 v_2 + v_1 \overline{u_2}).$$

Exercises

The complex conjugate $\overline{a+bi} = a - bi$ has some properties that lead elegantly to the multiplicative property of absolute value found in Section 2.5.

6.5.1 By setting $u = a + bi$ and $v = c + di$, calculate $u + v$, uv, $\overline{u+v}$ and $\overline{uv}$. Hence show that $\overline{u+v} = \overline{u} + \overline{v}$ and $\overline{u \cdot v} = \overline{u} \cdot \overline{v}$.

6.5.2 Show also that $|w|^2 = w \cdot \overline{w}$ for any $w = e + fi$.

6.5.3 Deduce from 6.5.1 and 6.5.2 that $|uv|^2 = |u|^2|v|^2$, so that $|uv| = |u||v|$. Also deduce the Diophantus identity

$$(a^2 + b^2)(c^2 + d^2) = (ac - bd)^2 + (bc + ad)^2.$$

When quaternions are defined as pairs $q = (u, v)$ of complex numbers, we use the addition rule $(u_1, v_1) + (u_2, v_2) = (u_1 + u_2, v_1 + v_2)$ and the above multiplication rule

$$(u_1, v_1)(u_2, v_2) = (u_1 u_2 - v_1 \overline{v_2}, u_1 v_2 + v_1 \overline{u_2}) \tag{*}$$

It can then be checked that all the laws of algebra hold *except* commutative multiplication.

6.5.4 Show using (*) that $(0, i)(i, 0) \neq (i, 0)(0, i)$.

6.6   Quaternions and Space Rotations

In Section 2.6 we saw that complex numbers can represent rotations of the plane. If u is a complex number with $|u| = 1$, then multiplication of the whole plane of complex numbers by u is a rotation about O that moves 1 to u. The number u can also be written

$$u = \cos\theta + i\sin\theta,$$

where θ is the angle between the directions of 1 and u (Figure 6.5).

Hence *rotation of the plane through angle θ is effected by multiplying by $\cos\theta + i\sin\theta$.*

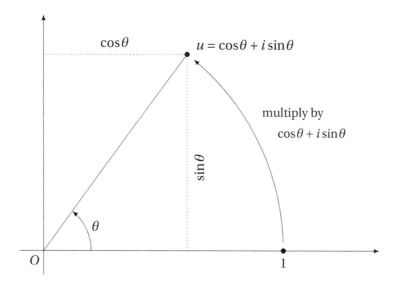

Figure 6.5: Rotation of the plane through angle θ.

We also saw that multiplying the whole four-dimensional space $\mathbb{R}^4$ of quaternions by a quaternion u with $|u| = 1$ is a rigid motion of $\mathbb{R}^4$ that leaves O fixed. Because of this, it is natural to use quaternions to study rotations of $\mathbb{R}^4$. However, it would be nice to understand rotations of three-dimensional space $\mathbb{R}^3$ first, and the pleasant surprise is that quaternions are good for this task (so much so, in fact, that they have become a standard tool in computer animation). The noncommutative multiplication is precisely what makes quaternions eligible to represent space rotations, because *space rotations do not generally commute.*

Here is an example. Take an equilateral triangle ABC placed in the plane of the paper, and consider the effect of two rotations on it:

- a 1/3 turn in the plane, clockwise, about the center of the triangle,

- a 1/2 turn in space, about a line through the top vertex of the triangle and its center.

Figure 6.6 shows the effects of combining these rotations: on the one hand with the 1/3 turn first, on the other hand with the 1/2 turn first.

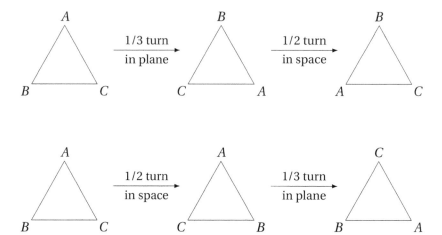

Figure 6.6: Combining rotations of the triangle.

As you can see from the final positions of the vertices A, B, C, the two combinations are different. Hence these two rotations do not commute.

Quaternion Representation of Space Rotations

To prepare the ground for quaternions we interpret each triple (x, y, z) in $\mathbb{R}^3$ as the *pure imaginary quaternion* $x\mathbf{i} + y\mathbf{j} + z\mathbf{k}$. Thus we are calling the coordinate axes in $\mathbb{R}^3$ the $\mathbf{i}$-, $\mathbf{j}$-, and $\mathbf{k}$-axes, and we call the space itself $(\mathbf{i}, \mathbf{j}, \mathbf{k})$-space. We also simplify our quaternion notation at this point, writing the identity quaternion $\mathbf{1}$ as 1, and omitting the factor 1 from products, so the typical quaternion is now written $a + b\mathbf{i} + c\mathbf{j} + d\mathbf{k}$. This makes the *real part a* more clearly distinct from the *imaginary part* $b\mathbf{i} + c\mathbf{j} + d\mathbf{k}$, and parallels the way we write complex numbers.

Now a rotation of the plane is given by an *angle* θ (the amount of turn) and a *center* (the fixed point, which we always choose to be O). Similarly, a rotation of three-dimensional space is specified by an angle θ and an *axis* (the line about which the turn through angle θ takes place). Each point of the axis is fixed, and we will be interested only in the case where the axis passes through O. In this case, the axis of rotation can be specified by either of the points at which it meets the unit sphere, and hence by a quaternion of the form $\lambda\mathbf{i} + \mu\mathbf{j} + \nu\mathbf{k}$, where $\lambda^2 + \mu^2 + \nu^2 = 1$.

The quaternion $\lambda\mathbf{i} + \mu\mathbf{j} + \nu\mathbf{k}$ specifying the rotation axis plays a similar role to that of the number i in the complex number $\cos\theta + i\sin\theta$ that rotates the plane through angle θ about O. In fact: *a space rotation through angle θ about axis $\lambda\mathbf{i} + \mu\mathbf{j} + \nu\mathbf{k}$ is effected by the quaternion*

$$u = \cos\frac{\theta}{2} + (\lambda\mathbf{i} + \mu\mathbf{j} + \nu\mathbf{k})\sin\frac{\theta}{2}.$$

At first sight, the $\theta/2$ looks like a mistake. How can this expression effect a rotation through angle θ? The explanation is that the quaternion u effects the rotation *not* simply by multiplication—this would not map $(\mathbf{i}, \mathbf{j}, \mathbf{k})$-space onto itself—but by a map called *conjugation*, which sends each pure imaginary quaternion q to uqu^{-1}. It can be checked that if q is a pure imaginary quaternion, so is uqu^{-1}.

This representation of rotations by quaternions was discovered independently by Cayley and Hamilton in 1845. Cayley also noticed that the same parameters θ, λ, μ, and ν had been used by Rodrigues in 1840, and that Rodrigues' rule for finding the parameters of the combination of two rotations was essentially the quaternion multiplication rule. When Gauss's unpublished work came to light, it was found that he too had essentially discovered quaternion multiplication.

Exercises

Consider the following two rotations:

- $r_x = 90°$ rotation about the x-axis that moves the y-axis to the z-axis.

- $r_y = 90°$ rotation about the y-axis that moves the z-axis to the x-axis.

6.6.1 Sketch a picture of the axes and use it to find where the point $(1, 0, 0)$ is moved by the combination of r_x, then r_y.

6.6.2 Similarly find where $(1, 0, 0)$ is moved by r_y, then r_x.

6.6.3 Does $r_x r_y = r_y r_x$?

6.7   Symmetry in Three Dimensions

One of the delights of three-dimensional space $\mathbb{R}^3$ is that it is the home of five highly symmetric objects called the *regular polyhedra*: the tetrahedron, cube, octahedron, dodecahedron, and icosahedron (Figure 6.7). Each regular polyhedron is a convex solid bounded by identical polygon faces. The tetrahedron, octahedron, and icosahedron are bounded respectively by four, eight, and 20 equilateral triangles; the cube by six squares; the dodecahedron by 12 regular pentagons. A regular polyhedron not only has regular polygon faces, but also regular vertices, in the sense that the same number of faces meet at each vertex: three at each vertex of the tetrahedron, cube, and dodecahedron, four at each vertex of the octahedron, and five at each vertex of the icosahedron.

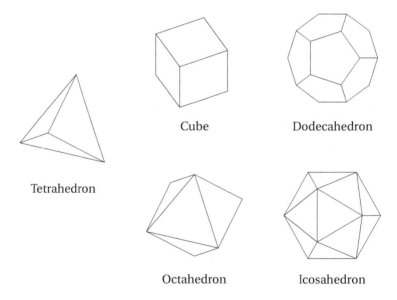

Cube   Dodecahedron

Tetrahedron

Octahedron   Icosahedron

Figure 6.7: The regular polyhedra.

By considering the size of the angles in regular polygons, it can be shown that the five polyhedra of Figure 6.7 are the only ones that are both convex and with regular faces and vertices. Also, it follows from these conditions that they have regular edges, in the sense that the angles between adjacent faces of a regular polyhedron are all the same (for example, all right angles for a cube). Thus the regular polyhedra are rare and precious jewels, and it is no surprise that geometers have admired them since ancient times. The climax of Euclid's *Elements* is a proof of the existence and uniqueness of the five regular polyhedra, and the Greeks may have been interested in irrational numbers because they occur in the regular polyhedra.

It is striking that only five regular polyhedra exist, because of course there are infinitely many regular polygons—a regular n-gon for each natural number $n \geq 3$. Three-dimensional symmetry is a rare phenomenon, because only the five regular polyhedra are fully symmetrical, in the sense that any face (or vertex, or edge) of a regular polyhedron looks the same as any other. Moreover, the five regular polyhedra have only *three types* of symmetry between them, because the cube and the octahedron have the *same* symmetry type, as do the dodecahedron and the icosahedron.

The reason that the cube and octahedron have the same symmetry type is shown in Figure 6.8, in which each vertex of the octahedron lies at a face center of the cube. A *symmetry of* any object is a motion that leaves the object looking the same; for example, one symmetry of the cube is a 1/4 turn about an axis through the centers of opposite faces. It is clear from the figure that each symmetry of the cube is a symmetry of the octahedron, and vice versa.

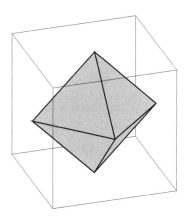

Figure 6.8: The octahedron and the cube.

We similarly see that the dodecahedron and the icosahedron have the same symmetry type by taking the vertices of the icosahedron to lie at the face centers of the dodecahedron. Thus the concept of "symmetry type" is an abstraction that can be made concrete in various ways. The cube and the octahedron are two objects that embody the same abstract symmetry type—they *crystallize* it, you might say—and we shall see in the next section that symmetry types can also be captured algebraically, by finite sets of quaternions.

The tetrahedron has a symmetry type all its own. It is in a sense "half as symmetric" as the cube, because a tetrahedron can be placed inside a cube in such a way (Figure 6.9) that all its symmetries are symmetries of the cube, but only half of the symmetries of the cube are symmetries of the tetrahedron. There are 12 symmetries of the cube that map the tetrahedron onto itself—we enumerate them in the next section—and 12 that do not.

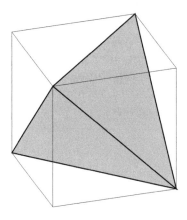

Figure 6.9: The tetrahedron and the cube.

Exercises

The regular polyhedra are related to the tilings of the sphere studied in the exercises to Sections 5.2 and 5.4. In particular, if we surround a regular polyhedron by a sphere with the same center as the polyhedron, and project the edges of the polyhedron onto the sphere from the center, then we obtain a tiling of the sphere by spherical polygons. Figure

6.10 shows this projection of the cube. (These pictures are courtesy of Henry Segerman, of Oklahoma State University.)

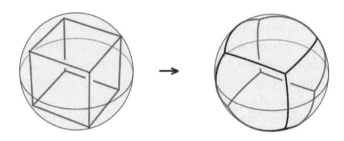

Figure 6.10: Projection of cube onto sphere.

6.7.1 Which polyhedron projects the tiling by triangles shown in Figure 5.15?

The tilings of the sphere by triangles seen in Section 5.2 are not projections of regular polyhedra, but *subdivisions* of such projections. In fact they come from projections of the tetrahedron, octahedron, and icosahedron by dividing the projection of each face into six triangles.

6.7.2 Which tiling comes from the tetrahedron?

6.7.3 Which tiling comes from the octahedron? Is this tiling also a subdivision of the cube tiling?

6.7.4 Which tiling comes from the icosahedron? Is this tiling also a subdivision of the dodecahedron tiling?

The theorem developed in the exercises to Section 5.4—

$$V - E + F = 2$$

when V, E, F are the numbers of vertices, edges, and faces in a tiling of the sphere—also applies to polyhedra. In fact, this is the reason it is called the Euler *polyhedron* formula.

6.7.5 Calculate V, E, and F for each regular polyhedron, and check in each case that $V - E + F = 2$.

6.8   Tetrahedral Symmetry and the 24-Cell

Our goal in this section is to describe the 12 symmetries of the tetrahedron as explicit rotations of $\mathbb{R}^3$, and then to examine the set of quaternions representing them as a set of points in $\mathbb{R}^4$. As we will see, the 12 rotations correspond to 24 quaternions—two for each rotation—forming a highly symmetric set of points. Joining each of the 24 points to its neighbors by edges, faces, and cells gives a four-dimensional object called the *24-cell*, which is of remarkable interest both geometrically and algebraically.

But first, why are there 12 symmetries of the tetrahedron? One way to see this is to choose a fixed position of the tetrahedron—a "tetrahedral hole" in space as it were—and then to count how many ways the tetrahedron can be slotted into it. Since any face of the tetrahedron is the same as any other, we can choose any one of the four faces to match a fixed face of the hole, say the *front* face. Each of the four faces that can go in front has three edges that can match a given edge, say the *bottom* edge, in the front face of the hole.

This gives 4 × 3 = 12 different ways in which the tetrahedron can occupy the same position, each corresponding to a different symmetry. But once we have chosen a particular face to go in front, and a particular edge of that face to go on the bottom, the symmetry is completely determined. Hence there are exactly 12 symmetries. Each can be obtained, from a given initial position, by a rotation. The rotations are about two kinds of axis, shown in Figure 6.11.

First there is the *trivial rotation*, which gives the *identity symmetry*, obtained by rotation through angle zero (about any axis). Then there are 11 nontrivial rotations, divided into two different types:

- The first type is a 1/2 turn about an axis through centers of opposite edges of the tetrahedron (also through opposite face centers of the cube). There are three such axes, and hence three rotations of this type.

- The second type is a 1/3 turn about an axis through a vertex and the face center opposite to it (which also goes through opposite vertices of the cube). There are four such axes, and hence *eight* rotations of this type—since the 1/3 turn clockwise is different from the 1/3 turn anticlockwise.

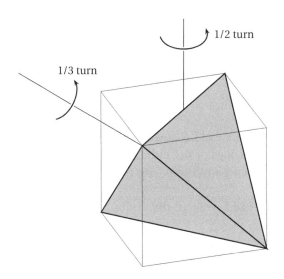

Figure 6.11: The tetrahedron and axes of rotation.

Notice also that each 1/2 turn moves all four vertices, whereas each 1/3 turn leaves one vertex fixed and moves the remaining three. Thus the 11 nontrivial rotations are all different and therefore, together with the trivial rotation, they account for all 12 symmetries of the tetrahedron.

Representation of Tetrahedron Rotations by Quaternions

As mentioned in Section 6.7, a rotation of $(\mathbf{i}, \mathbf{j}, \mathbf{k})$-space through angle θ about axis $\lambda\mathbf{i} + \mu\mathbf{j} + \nu\mathbf{k}$ corresponds to the quaternion

$$\cos\frac{\theta}{2} + (\lambda\mathbf{i} + \mu\mathbf{j} + \nu\mathbf{k})\sin\frac{\theta}{2}.$$

If we choose coordinate axes so that the sides of the cube in Figure 6.11 are parallel to the $\mathbf{i}$, $\mathbf{j}$, and $\mathbf{k}$ axes, then the axes of rotation are virtually immediate, and the corresponding quaternions are easy to work out.

- We can take the lines through opposite face centers of the cube to be the $\mathbf{i}$, $\mathbf{j}$, and $\mathbf{k}$ axes. For a 1/2 turn, the angle $\theta = \pi$, and hence $\theta/2 = \pi/2$. Therefore, since $\cos\frac{\pi}{2} = 0$ and $\sin\frac{\pi}{2} = 1$, the 1/2 turns about the $\mathbf{i}$, $\mathbf{j}$, and $\mathbf{k}$ axes are given by the quaternions $\mathbf{i}$, $\mathbf{j}$, and $\mathbf{k}$ themselves.

However, if the **i** axis is an axis of rotation, so is its "other half," the −**i** axis. Hence the 1/2 turn about this axis is also represented by the quaternion −**i**. Thus the three 1/2 turns are represented by the three pairs of opposites

$$\pm\mathbf{i}, \quad \pm\mathbf{j}, \quad \pm\mathbf{k}.$$

- Given the choice of **i**, **j**, and **k** axes, the four rotation axes through opposite vertices of the cube correspond to four pairs of opposites, which together make up the eight combinations

$$\frac{1}{\sqrt{3}}(\pm\mathbf{i}\pm\mathbf{j}\pm\mathbf{k}) \quad \text{(independent choices of + or − sign).}$$

The factor $\frac{1}{\sqrt{3}}$ is to give each of these quaternions the absolute value 1, as specified for the representation of rotations.

For each 1/3 turn we have $\theta = \pm 2\pi/3$, and hence

$$\cos\frac{\theta}{2} = \cos\frac{\pi}{3} = \frac{1}{2}, \qquad \sin\frac{\theta}{2} = \pm\sin\frac{\pi}{3} = \pm\frac{\sqrt{3}}{2}.$$

The values of $\cos\frac{\pi}{3}$ and $\sin\frac{\pi}{3}$ can be seen in the right-angled triangle which is half an equilateral triangle (Figure 6.12).

The $\sqrt{3}$ in $\sin\frac{\pi}{3}$ neatly cancels the factor $1/\sqrt{3}$ in the axis of rotation, and we find that the eight 1/3 turns are represented by the eight pairs of opposites among the sixteen quaternions

$$\pm\frac{1}{2}\pm\frac{\mathbf{i}}{2}\pm\frac{\mathbf{j}}{2}\pm\frac{\mathbf{k}}{2}.$$

Finally, the identity rotation is represented by the pair ±1, and thus the 12 symmetries of the tetrahedron are represented by the 24 quaternions

$$\pm 1, \quad \pm\mathbf{i}, \quad \pm\mathbf{j}, \quad \pm\mathbf{k}, \quad \pm\frac{1}{2}\pm\frac{\mathbf{i}}{2}\pm\frac{\mathbf{j}}{2}\pm\frac{\mathbf{k}}{2}.$$

The 24-Cell

These 24 quaternions all lie at distance 1 from O in $\mathbb{R}^4$, and they are distributed in a highly symmetrical manner. In fact, they are the vertices of a four-dimensional figure analogous to a regular polyhedron— a *regular polytope*. This particular polytope is called the *24-cell*. Its

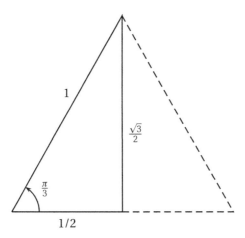

Figure 6.12: Why $\cos \frac{\pi}{3} = 1/2$ and $\sin \frac{\pi}{3} = \sqrt{3}/2$.

edges are the line segments joining each vertex to its nearest neighbors, its *faces* are triangles spanning triples of vertices joined in pairs, and these faces span 24 octahedra, which bound the 24-cell in the same sense that faces bound a polyhedron and edges bound a polygon. (The names "polygon," "polyhedron," and "polytope" say, roughly, what the boundaries are made of: they come from the Greek for "many corners," "many faces," and "many cells," respectively.)

There are just six regular polytopes in $\mathbb{R}^4$, and in the next section we shall explain where the 24-cell fits among them.

Exercises

6.8.1 By taking **i, j, k** as unit vectors along the x, y, z axes, respectively, find quaternions that represent the rotations r_x and r_y described above.

6.8.2 Use quaternions to decide whether $r_x r_y = r_y r_x$.

6.8.3 Show that the point $\frac{1}{2} + \frac{\mathbf{i}}{2} + \frac{\mathbf{j}}{2} + \frac{\mathbf{k}}{2}$ is at distance 1 from O in $\mathbb{R}^2$.

6.8.4 Deduce from Question 6.8.3 that the distance from the center of a 4-cube to any vertex equals the length of a side.

It appears from Figure 6.15 that each vertex of the 24-cell has eight nearest neighbors (this is clearest for the bottom vertex).

6.8.5 Show that 1 is the smallest distance between any two of the 24 quaternion vertices of the 24-cell.

6.8.6 Also find eight vertices at distance 1 from the vertex **i**.

6.9   The Regular Polytopes

We cannot directly visualize four-dimensional polytopes, but we can know them to a large extent by their "shadows" in three-dimensional space. This is analogous to the way we normally learn about polyhedra. We are quite used to perceiving three-dimensional objects through two-dimensional images—after all, this is exactly what our eyes do. Regular polyhedra can certainly be recognized from their shadows, as one sees from the shadows of the tetrahedron, cube, and octahedron shown in Figure 6.13. These shadows are cast by a point source of light placed outside a wire frame polyhedron, near to a face center. Equivalently, they are what you see when looking into the polyhedron through one face from close range—near enough to see all the other faces through it.

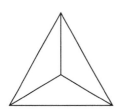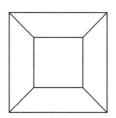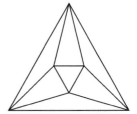

Figure 6.13: Shadows of the tetrahedron, cube, and octahedron.

The tetrahedron shadow, for example, is a big triangle with three smaller triangles inside it. These four triangles are the shadows of the four faces of the tetrahedron.

There are analogues of these three polyhedra in all dimensions, and those in four-dimensional space are called the *4-simplex, 4-cube,* and *4-orthoplex*, respectively. Their shadows in three dimensions are shown in Figure 6.14. Of course, these pictures are really planar, but we read them as frameworks in three-dimensional space. The framework is what a four-dimensional being would see when looking into the polytope through one cell from close range.

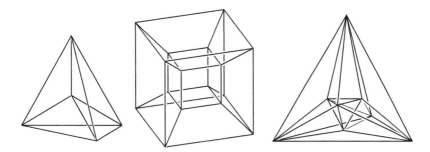

Figure 6.14: Shadows of the 4-simplex, 4-cube, and 4-orthoplex.

The rods in the framework outline *cells* that are shadows of boundary cells of the polytope. The 4-simplex shadow, for example, is a big tetrahedron with four smaller tetrahedra inside it. These five tetrahedra are the shadows of the five boundary tetrahedra of the 4-simplex. The 4-cube and the 4-orthoplex are *dual* to each other, like the cube and octahedron. That is, the cell centers of the 4-cube are vertices of a 4-orthoplex, and vice versa.

A shadow of the 24-cell is shown in Figure 6.15, a picture taken from the classic book *Geometry and the Imagination* by Hilbert and Cohn-Vossen. It consists of a big octahedron with 23 small octahedra inside it—the shadows of the 24 boundary octahedra of the 24-cell. The 24-cell has no counterpart in three-dimensional space, nor in any dimension higher than four.

Indeed, there are only three regular polytopes in each $\mathbb{R}^n$ for $n > 4$: namely, the analogues of the tetrahedron, cube, and octahedron. (The word "polytope" is used for polyhedron analogues in all dimensions from four upwards.) And there are only five *other* regular polyhedra and polytopes: in $\mathbb{R}^3$ the dodecahedron and icosahedron, and in $\mathbb{R}^4$ the 24-cell and two others, called the *120-cell* and the *600-cell*. As their names suggest, the latter polytopes are bounded by 120 cells (which happen to be dodecahedra) and 600 cells (which happen to be tetrahe-

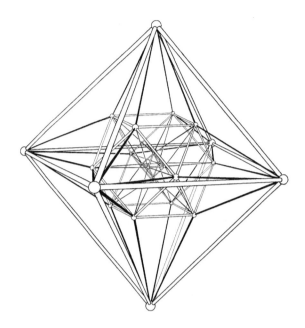

Figure 6.15: Shadow of the 24-cell.

dra), respectively. The 120-cell and the 600-cell are dual to each other, and the 120 vertices of the 600-cell correspond in pairs to the symmetries of the icosahedron. It seems almost impossible to produce really clear diagrams of the 120-cell and 600-cell—models of their three-dimensional shadows are better—but for some interesting attempts see the book *Regular Polytopes* by H. S. M. Coxeter or my article *The Story of the 120-Cell* at

www.ams.org/notices/200101/fea-stillwell.pdf.

The regular polytopes in all dimensions were found by the Swiss mathematician Ludwig Schläfli in 1852. In particular, Schläfli discovered the 24-cell, 120-cell, and 600-cell, and proved that they are the only regular polytopes other than the higher-dimensional analogues of the tetrahedron, cube, and octahedron. He also found all the *regular space tilings* in dimensions greater than three. These are analogous to the tilings of the plane by squares, equilateral triangles, and hexagons shown in Figure 6.16, and the tiling of $\mathbb{R}^3$ by cubes shown in Figure 5.1.

The tilings of the plane by triangles and hexagons are exceptional, because they have no analogues in higher dimensions, and they are *dual* to each other in the sense that the vertices of one tiling are the face centers of the other.

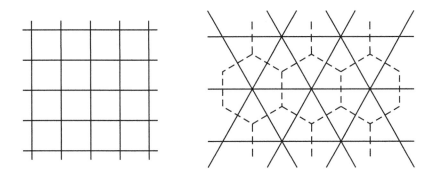

Figure 6.16: Regular tilings of the plane.

The square and cube tilings generalize to a regular tiling of $\mathbb{R}^n$ by n-cubes. Its vertices are the n-tuples of integers. But apart from this "obvious" tiling in each $\mathbb{R}^n$ there are only two regular space tilings, both in dimension four. One of them divides $\mathbb{R}^4$ into 24-cells and the other is its dual, obtained by replacing each 24-cell by the point at its center. The resulting points are the vertices of a regular tiling of $\mathbb{R}^4$ by 4-orthoplexes. With suitable choice of coordinate axes, the vertices of the orthoplex tiling are the quaternions of the form

$$a\left(\frac{1}{2} + \frac{\mathbf{i}}{2} + \frac{\mathbf{j}}{2} + \frac{\mathbf{k}}{2}\right) + b\mathbf{i} + c\mathbf{j} + d\mathbf{k} \quad \text{for integers } a, b, c, d.$$

Exercises

Since the vertices of the 24-cell are at distance 1 from the origin in 4-dimensional space, they lie on the 3-sphere of *all* points at distance 1 from the origin. We can then consider a "spherical 24-cell" (like a spherical triangle) whose edges join vertices of the 24-cell by great circle arcs rather than straight lines. Thus the "spherical 24-cell" lies entirely in the 3-sphere. Then we can *stereographically project* the 3-sphere into an ordinary 3-dimensional space, as we did in Section 5.2.

As we know from Section 5.2, stereographic projection sends circles to circles, so we get an image of the 24-cell whose edges are circular arcs. Also, its faces are portions of spheres. Projecting this image, in turn, onto the plane of the page, we get Figure 6.17.

This figure was created by Fritz Obermeyer, and is available through Wikimedia at

https://commons.wikimedia.org/wiki/
File:Stereographic_polytope_24cell_faces.png.

Figure 6.17: Stereographic projection of the spherical 24-cell.

6.9.1 Check by counting that the outer surface of this image consists of eight faces that are parts of spheres.

6.9.2 Assuming that the angles at each vertex are equal, what is the angle?

6.9.3 Conclude, from what you know about spherical triangles, that the outer faces are *not* spherical triangles, hence their edges are not great circles on the parts of spheres that they bound.

Chapter 7

The Ideal

Preview

In this chapter we return to the natural numbers 1, 2, 3, 4, 5, ... with which we started in Chapter 1. Since then we have seen how the number concept expanded in response to various demands and crises, but we have not yet seen many properties of the natural numbers themselves.

The natural numbers and integers differ from the real numbers in having "atoms" that cannot be "split" into products. These are the *primes*, and they are the key to many properties of natural numbers because of *unique prime factorization*: each number > 1 is the product of primes in only one way.

Unique prime factorization is (essentially) in Euclid's *Elements*. But little else was known about primes until 1640, when Fermat announced some startling theorems about primes and solutions of equations. Over a century later, Euler and Gauss explained Fermat's theorems by "splitting" integers into *complex integers*, for example, splitting $x^2 + 2$ into $(x + \sqrt{-2})(x - \sqrt{-2})$.

However, complex integers are really helpful only when they split uniquely into complex primes. There are complex integers *without* this property, such as the numbers $a + b\sqrt{-5}$ where a and b are ordinary integers. *So at this point it seems impossible to enjoy the advantages of unique prime factorization any longer.* The German mathematician Ernst Eduard Kummer discovered this obstacle in the 1840s

207

and, quixotically, he proposed to defy it.

Kummer believed he could cure badly behaved primes by splitting them into what he called "ideal primes." The "ideal primes" he yearned for were not even known to exist when he first used them! We explain how they may be found as *greatest common divisors* of complex primes, which ties up nicely with Euclid's approach to the ordinary primes.

7.1 Discovery and Invention

We speak of invention: it would be more correct to speak of discovery. The distinction between these two words is well known: discovery concerns a phenomenon, a law, a being which already existed, but had not been perceived. Columbus discovered America: it existed before him; on the contrary, Franklin invented the lightning rod: before him there had never been any lightning rod.

Such a distinction has proved less evident than it appears at first glance. Torricelli has observed that when one inverts a closed tube on a mercury trough, the mercury ascends to a certain determinate height: this is a discovery; but, in doing this, he has invented the barometer; and there are plenty of examples of scientific results which are just as much discoveries as inventions.

Jacques Hadamard, *An Essay on the Psychology of Invention in the Mathematical Field*, p. xi.

Hadamard's remarks on invention versus discovery strike a chord with many mathematicians, but what is the difference between invention and discovery in mathematics? Briefly, mathematicians generally believe that mathematical results are discovered, but language, notation, and proofs are invented to discuss and communicate them. These human inventions are far from the perfect embodiment of mathematical facts; rather, they are an ad hoc and temporary means of accessing them. Grasping a mathematical fact via words and symbols can be hard, but once we have "got" it, the fact becomes virtually independent of language. Pictures can help. We easily recognize the Pythagorean theorem when we see it in a Chinese or Indian book, even if the language is unknown to us, and we recognize that hundreds of different proofs (collected in the book of Loomis, *The Pythagorean Proposition*) are all proofs of this same theorem.

Much could be said about language, thought, and communication, and their influence on invention and discovery. But in this chapter we focus on just one mathematical fact—unique prime factorization—that illustrates the interplay between discovery and invention in a particularly striking way.

First we restrict attention to the natural numbers or *positive integers* 1, 2, 3, 4, 5, ... and among them the *primes*: those integers greater than 1 that are not products of smaller positive integers. Thus the first few primes are

$$2, 3, 5, 7, 11, 13, 17, 19, 23, 29, 31, 37, 41, 43, 47, 53, 59, 61, 67, 71, 73, 79, \ldots$$

In a sense, the concept of prime is a linguistic invention, an abbreviation for a particular type of integer. But the more we discover about primes the more useful the abbreviation becomes. Linguistic inventions may be arbitrary but they are subject to a kind of natural selection. Only the fittest inventions survive, and primes have survived so far because they best express some fundamental properties of numbers.

It follows easily from the definition of prime that *every positive integer greater than 1 is a product of primes*. If the number n is not itself a prime, then it is a product of smaller numbers greater than 1, a and b say. If a or b is not itself a prime, it too can be written as a product of smaller numbers greater than 1, and so on. Since positive integers cannot decrease forever, it takes only a finite number of steps to write n as a product of primes.

For example, 60 can be written as the product 4×15, and the 4 and 15 break down into products of primes: $4 = 2 \times 2$ and $15 = 3 \times 5$. Hence a prime factorization of 60 is

$$60 = 2 \times 2 \times 3 \times 5.$$

We do not claim, yet, that this is the only prime factorization of 60. The number 60 has many different factorizations, for example,

$$60 = 4 \times 15 = 6 \times 10 = 5 \times 12,$$

and we have not checked that 6×10 and 5×12 break down into the same product of primes. (Try this for yourself.)

It so happens that *all* factorizations of 60 break down to $2 \times 2 \times 3 \times 5$, but there is no guarantee that a similar thing will happen with other numbers. In general, we have proved only the *existence* of prime factorization, which is often expressed by saying that the primes are the multiplicative "building blocks" for the positive integers.

What we really want is *uniqueness* of prime factorization, which says that each number can be built from primes in only one way: *every positive integer greater than 1 can be written uniquely as a product of primes* (in nondecreasing order). We defer the proof of unique prime factorization until later, and give only some examples:

$$30 = 2 \times 3 \times 5, \qquad 60 = 2 \times 2 \times 3 \times 5, \qquad 999 = 3 \times 3 \times 3 \times 37.$$

No matter how you break down any of these numbers into smaller factors, you will find that the ultimate factors are the primes given here. The result is much subtler than existence of prime factorization, because the breakdown from large factors to small can be made in many different ways, and there is no obvious reason why it should always lead to the same result.

Remark. Since 1 is not a prime, it does not have a prime factorization. Is this a flaw in our definition of prime? I do not mean, of course, that the definition is wrong; only that another definition may be more useful. The question is: is it more useful to call 1 a prime?

It does not affect the existence of prime factorization if 1 is prime; in fact the theorem is slightly beautified since 1 is no longer an exception. But *uniqueness* of factorization is spoiled when factors of 1 are allowed, since they can be present any number of times without affecting the value of the product. As you will see, we very much want unique prime factorization, so it is useful not to call 1 a prime. (At one time it was common to define the prime numbers so as to include 1, but natural selection—through the demand for unique factorization—has now made the old definition extinct.)

Exercises

7.1.1 Factorize the numbers 600, 601, and 602 into primes. Which two of 600, 601, 602 have a common prime factor?

7.1.2 Also factorize the number 603 into primes. Do any of 601, 602, and 603 have a prime factor in common?

7.1.3 Without attempting any factorization, explain why 10000 and 10002 have exactly one prime factor in common.

7.2   Division with Remainder

To discuss primes we need to be clear about the concept of *division* in the positive integers. This involves a return to the concept of division you learned in school: *division with remainder*. First we discuss divisibility, which is the relation between positive integers where one divides the other exactly.

Divisibility

We say that a positive integer b *divides* a if

$$a = qb \quad \text{for some positive integer } q.$$

Other common ways to express this relationship are: a is *divisible by b*, or a is a *multiple of b*. Some examples are: 6 divides 12, but 6 does not divide 13, 14, 15, 16, or 17. Divisibility is an interesting relationship because it occurs quite rarely and is hard to recognize. It is hard to see, for example, whether 321793 divides 165363173929, and even harder to see whether 165363173929 is divisible by *any* positive integer smaller than itself (except 1)—that is, it is hard to recognize whether 165363173929 is prime.

   This is one of the reasons for the interest in primes, though perhaps the basic reason is that the primes form such a mysterious and irregular sequence. There seems to be no pattern in the sequence except its *avoidance* of certain patterns: after the number 2 it includes no other multiple of 2; after 3, no other multiple of 3; after 5, no other multiple of 5; and so on. It is not even clear whether the sequence of primes is infinite, but in fact it is—and the proof of this fact by Euclid was one of the first successes for the concept of divisibility.

   Euclid's proof that there are infinitely many primes depends on a fact alluded to in Section 1.6: if c divides a and c divides b, then c divides $a - b$. As we saw there, if c divides a and b, then

$$a = mc \quad b = nc \quad \text{for some positive integers } m \text{ and } n,$$

and therefore

$$a - b = (m - n)c, \quad \text{so } c \text{ divides } a - b.$$

We now use this simple fact to show that, *given any primes $p_1, p_2, \ldots, p_n$ we can find another prime p*, and hence there are infinitely many primes.

Euclid's idea is in the *Elements*, Book IX, Proposition 20, and it is to consider $k = p_1 p_2 \cdots p_n$ and $k + 1$. Obviously, all of the primes $p_1, p_2, \ldots, p_n$ divide k, but if one of them divides $k + 1$ it will also divide the difference $k + 1 - k = 1$, which is impossible.

But *some* prime p divides $k + 1$, because any integer > 2 is a product of primes, as we saw in the preceding section. Thus p is a prime number different from the given primes $p_1, p_2, \ldots, p_n$. $\square$

Remainder

Since it is not always true that b divides a, we need the more general concept of division with remainder. As you know from carrying out divisions in school, division of a by b results in a *quotient q* and a *remainder r*. Exact division occurs only when $a = qb$ and $r = 0$, but what can we say in general about the size of r? The general situation is that a falls between two successive multiples of b, qb and $(q + 1)b$, as shown in Figure 7.1.

Figure 7.1: Integer multiples of b and the remainder r.

From this it is clear that $a = qb + r$, where $0 \le r < b$. As long as $b \ne 0$, it is also clear that we can drop the assumption that a and b are positive, because it is still true that a lies between some qb and $(q+1)b$. This gives the *division property of integers: for any integers a and $b \ne 0$ there are integers q and $r \ge 0$ such that $a = qb + r$, where $0 \le r < |b|$.*

There are better ways of finding the numbers q and r than writing down all these multiples of b, and Figure 7.1 is not the only way to see the *existence* of r less than b either. But when we generalize the concept of integer to complex numbers, a two-dimensional version of this geometric argument is by far the easiest way to see whether or not a small remainder exists. (Sometimes it doesn't, and this leads to some surprises.)

The Greatest Common Divisor

The idea of displaying multiples of an integer as equally spaced points along the number line gives a striking insight into the *greatest common divisor* of integers a and b, the greatest integer that divides both a and b. We begin by showing what happens when $a = 6$ and $b = 8$.

First, here are the multiples $6m$ of 6:

Next, here are the multiples $8n$ of 8:

And finally, here are all the sums of these multiples, $6m + 8n$:

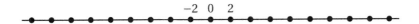

It is surprising, at first, that the numbers $6m + 8n$ are precisely the integer multiples of 2, the greatest common divisor of 6 and 8. (It depends very much on the presence of negative m and n: if we allow only positive m and n, then the numbers $6m + 8n$ form the irregular sequence $6, 8, 12, 14, 16, 18, \ldots$.) A little thought uncovers the reasons for this result:

- Since 6 is a multiple of 2 and 8 is a multiple of 2, every number of the form $6m + 8n$ is a multiple of 2.

- The number 2 equals $-6+8$, hence any integer multiple of 2 is of the form $2q = -6q + 8q$, which is of the form $6m + 8n$.

- Hence the numbers $6m+8n$ are precisely the multiples of 2, which is the greatest common divisor of 6 and 8.

A similar argument lets us describe the greatest common divisor, $\gcd(a, b)$, of any integers a and b: *the integers of the form $ma + nb$, for integers m and n, are precisely the integer multiples of* $\gcd(a, b)$.

The reasons are like those uncovered in the special case above, except that division with remainder now makes an appearance (it went unnoticed in the example, because the gcd of 6 and 8 is clearly 2).

- Since a is a multiple of $\gcd(a, b)$ and b is a multiple of $\gcd(a, b)$, every number of the form $ma + nb$ is a multiple of $\gcd(a, b)$.

- Let c be the *smallest* positive value of $ma + nb$. Then all integer multiples of c are also of the form $ma + nb$. Conversely, every value of $ma + nb$ is a multiple of c. Why? If some $m'a + n'b$ is *not* a multiple of c, consider its remainder, $m'a + n'b - qc$ on division by c. Since qc is of the form $ma + nb$, the remainder also has this form. But the division property says that the remainder is smaller than c, and a positive number $ma + nb$ smaller than c contradicts the definition of c.

- Thus the numbers $ma + nb$ are precisely the multiples of their smallest positive member, c. The numbers $ma + nb$ include a and b, hence a and b are multiples of c; that is, c is a common divisor of a and b. But we already know that $\gcd(a, b)$ divides all numbers $ma + nb$, hence $\gcd(a, b)$ divides c and therefore $c = \gcd(a, b)$.

Exercises

For small numbers a and b, $\gcd(a, b)$ can be found by looking at the prime factors of a and b (and assuming unique prime factorization). For example,

$$30 = 2 \times 3 \times 5,$$
$$63 = 3^2 \times 7,$$

hence $\gcd(30, 63) = 3$.

7.2.1 Find $\gcd(60, 98)$ by factorizing 60 and 98.

7.2.2 Similarly, find $\gcd(55, 34)$.

Another quantity that is easily computed from prime factorizations is the *least common multiple*, $\text{lcm}(a, b)$.

7.2.3 Explain why $\text{lcm}(30, 63) = 2 \times 3^2 \times 5 \times 7$.

7.2.4 Find $\text{lcm}(55, 34)$.

7.3 The Euclidean Algorithm

The results about the gcd in the previous section are fine in theory, but not much help if we want to find the gcd of two large numbers—perhaps even the example of 10000 and 10002 in Exercise 7.1.3. For this and larger examples a faster and simpler approach is desirable. It depends on the simple fact, discussed in the previous section, that *any common divisor of a and b is also a divisor of a − b*. This makes quick work of $a = 10002$ and $b = 10000$, because it tells us that their common divisors also divide 2, so 2 is in fact the *greatest* common divisor of 10000 and 10002.

The presence of $\gcd(a, b)$ in $a - b$ is the basis of the *Euclidean algorithm* which, as Euclid put it, "repeatedly subtracts the lesser number from the greater." In other words, to find the gcd of two positive integers a and b, with $a > b$, replace them by the pair b and $a - b$, and repeat until two equal numbers are found.

Since the pairs continually decrease in size, and positive integers cannot decrease forever, this process eventually halts. And since all common divisors of a and b are retained at each step, the number ultimately found is $\gcd(a, b)$. For example, if we begin with the pair $a = 55$, $b = 34$ the algorithm begins by giving

$$\gcd(55, 34) = \gcd(34, 21) = \gcd(21, 13) = \cdots$$

and continues until we reach $\ldots \gcd(3, 2) = \gcd(2, 1) = \gcd(1, 1) = 1$.

Why and How $\gcd(a, b) = ma + nb$

The Euclidean algorithm gives another explanation why $\gcd(a, b)$ is of the form $ma + nb$ for some integers m and n. Namely,

- the initial numbers a and b are of this form,

- the difference between any two numbers of this form is of this form,

- in particular, the final number $\gcd(a, b)$ is of this form.

Moreover, we can enhance the Euclidean algorithm to *find* m and n such that $\gcd(a, b) = ma + nb$ for given a and b. The idea is to run the Euclidean algorithm simultaneously on the numbers represented by a and b and on the *actual letters* a and b, doing exactly the same operations on the letters a and b that we perform on the numbers they represent.

By operating on letters we can keep a record of the subtractions performed on the numbers, which the numbers alone cannot do. Thus, the number 5 does not tell us that it came from subtracting 8 from 13, but the expression $a - b$ *does* tell us that it came from subtracting b from a. Here is the record of subtractions when we compute $\gcd(13, 8)$.

$$
\begin{aligned}
\gcd(13, 8) &= \gcd(8, 13 - 8) & \gcd(a, b) &= \gcd(b, a - b) \\
&= \gcd(8, 5) & &= \gcd(b, a - b) \\
&= \gcd(5, 8 - 5) & &= \gcd(a - b, b - (a - b)) \\
&= \gcd(5, 3) & &= \gcd(a - b, -a + 2b) \\
&= \gcd(3, 5 - 3) & &= \gcd(-a + 2b, a - b - (-a + 2b)) \\
&= \gcd(3, 2) & &= \gcd(-a + 2b, 2a - 3b) \\
&= \gcd(2, 3 - 2) & &= \gcd(2a - 3b, -a + 2b - (2a - 3b)) \\
&= \gcd(2, 1) & &= \gcd(2a - 3b, -3a + 5b).
\end{aligned}
$$

There is now no need to go further, because the gcd 1 of 13 and 8 has appeared in the calculation with numbers, and its corresponding expression in the calculation with letters is $-3a + 5b$. So when $a = 13$ and $b = 8$ we have found

$$\gcd(a, b) = -3a + 5b,$$

and we can see that this is correct, because $-3 \times 13 + 5 \times 8 = 1$.

Using Division with Remainder Instead of Subtraction

Finding the gcd by repeated subtraction may be better than using prime factorizations, but it can still be absurdly inefficient. For example, it would be foolish to compute gcd(10001, 2) by subtracting 2 repeatedly from 10001 until the difference becomes 1. The efficient thing to do would be to *divide* 10001 by 2 and observe that the remainder is 1. Division with remainder for natural numbers yields the same result as repeated subtraction, so it likewise gives the gcd.

However, division with remainder is important for a reason other than speed: it works to give the gcd in settings where repeated subtraction does not. This is the case for the "complex integers" A and B we will meet later in this chapter, where division with remainder can*not* usually be achieved by repeated subtraction. In such cases we use division with remainder to prove the crucial result

$$\gcd(A, B) = MA + NB$$

for some "complex integers" M and N. This result is our stepping stone to unique prime factorization, as we will see in the next section.

Exercises

7.3.1 Use the Euclidean algorithm to find gcd(466, 288).

7.3.2 Find integers m and n such that $21m + 13n = 1$.

We can see directly that repeated division with remainder yields the gcd, via the Euclidean algorithm. It depends on the *division property* discussed in Section 7.2.

7.3.3 Explain why, if $a = bq + r$ where a, b, q, r are integers, that any divisor d of a and b is also a divisor of r.

7.3.4 Explain also, using Section 7.2 to justify taking $0 \le r < |b|$, that the process of replacing the pair a, b by the pair b, r eventually halts.

7.4   Unique Prime Factorization

The moral of the preceding sections is that divisors and primes may be a deep mystery, but *common* divisors are an open book. We have no simple formula for the divisors of an arbitrary integer a, but the gcd of a and b is just the smallest positive integer of the form $ma + nb$. It looks as though we should use the gcd to learn about primes. But how do we establish a connection between the two concepts?

One way is via the following observation: *if p is a prime and a is any integer not divisible by p, then* $\gcd(a, p) = 1$. This is true because the only positive divisors of p are p and 1, and of these two divisors only 1 divides a. The observation yields a fundamental property of prime numbers, given in Euclid's *Elements*, Book VII, Proposition 30 (though Euclid's proof is different):

Prime divisor property. *If a prime p divides the product ab of integers a and b, then p divides a or p divides b.*

To prove this, suppose that p does not divide a (thus we hope to show that p divides b). Then it follows from the observation above that $\gcd(a, p) = 1$. We know from the preceding section that $\gcd(a, p) = ma + np$ for some integers m and n, and therefore

$$1 = ma + np.$$

Now, to exploit the fact that p divides ab, we multiply both sides of this equation by b. This gives

$$b = mab + npb,$$

and we see that p divides both terms on the right side—it obviously divides npb, and it divides ab by assumption. Therefore p divides the sum of mab and npb, namely b, as required.   □

It follows from the prime divisor property that:

- If a prime p divides the product of positive integers $q_1, q_2, \ldots, q_s$, then p divides one of $q_1, q_2, \ldots, q_s$.

- If, also, the $q_1, q_2, \ldots, q_s$ are primes, then p equals one of them (because the only prime divisor of a prime is the prime itself).

- If $p_1 p_2 \cdots p_r$ and $q_1 q_2 \cdots q_s$ are equal products of primes, then each prime among $p_1, p_2, \ldots, p_r$ equals one of the primes among $q_1, q_2, \ldots, q_s$ (since each p_i divides $p_1 p_2 \cdots p_r$, which is equal to $q_1 q_2 \cdots q_s$) and vice versa.

Thus equal products of primes actually consist of the same prime factors, and we have:

Unique prime factorization. *Each positive integer > 1 is a product of primes in only one way, up to the order of factors.*

Unique prime factorization is a relatively modern theorem, first stated by Gauss in 1801, yet its core is the prime divisor property known to Euclid. One reason for the slow emergence of unique prime factorization may be the sophisticated notation required to discuss it. Euclid did not like to use more than three or four unknowns, let alone the "unknown number of unknowns" needed for the proof above. However, it is also true that the prime divisor property is often what one really needs. We shall see examples of this in later sections.

Irrational Roots of 2

In Chapter 1 we proved that $\sqrt{2}$ is irrational, and explained why this thwarts the Pythagorean program of dividing the octave into 12 equal intervals using whole number frequency ratios. We now use unique prime factorization to prove that $\sqrt[k]{2}$ *is irrational for any integer $k \geq 2$* (which prevents the octave being divided into *any* number k of equal intervals using whole number frequency ratios).

The proof begins by supposing (for the sake of contradiction) that

$$\sqrt[k]{2} = \frac{m}{n} \quad \text{for some integers } m \text{ and } n,$$

and that m and n have the prime factorizations

$$m = p_1 p_2 \cdots p_r, \quad n = q_1 q_2 \cdots q_s.$$

It follows, by raising both sides of $\sqrt[k]{2} = m/n$ to the power k, that

$$2 = (p_1 p_2 \cdots p_r)^k / (q_1 q_2 \cdots q_s)^k,$$

and therefore
$$2(q_1 q_2 \cdots q_s)^k = (p_1 p_2 \cdots p_r)^k.$$

This contradicts unique prime factorization! The number of occurrences of the prime 2 on the left is 1 plus some multiple of k (the multiple being the number of times 2 occurs among $q_1, q_2, \ldots, q_s$), whereas the number of occurrences of 2 on the right is a multiple of k.

Hence our original assumption, that $\sqrt[k]{2} = m/n$, is false.   □

Exercises

Suppose that $\sqrt{3} = m/n$ for some positive integers m and n, so $3 = m^2/n^2$ and hence $3n^2 = m^2$.

7.4.1 Explain why the factor 3 occurs an odd number of times in the prime factorization of $3n^2$, but an even number of times in the prime factorization of m^2.

7.4.2 Deduce from Exercise 7.4.1 that $\sqrt{3}$ is irrational.

7.4.3 Give a similar proof that $\sqrt{5}$ is irrational.

7.5   Gaussian Integers

There are many ways to use unique prime factorization, and it is rightly regarded as a powerful idea in number theory. In fact, it is more powerful than Euclid could have imagined. There are *complex* numbers that behave like "integers" and "primes," and unique prime factorization holds for them as well. Complex integers were first used around 1770 by Euler, who found they have almost magical powers to unlock secrets of ordinary integers. For example, by using numbers of the form $a + b\sqrt{-2}$ where a and b are integers, he was able to prove a claim of Fermat that 27 is the only cube that exceeds a square by 2. Euler's results were correct, but partly by good luck. He did not really understand complex "primes" and their behavior.

Gauss was the first to put the study of complex integers on a sound footing, in 1832. He studied what are now called the *Gaussian integers*: numbers of the form $a + bi$ where a and b are ordinary integers. Among them he found the "prime" Gaussian integers—now called *Gaussian primes*—and proved a unique factorization theorem for them. The ordinary integers are included among the Gaussian integers, but ordinary primes are not always Gaussian primes. For example, 2 has the "smaller" Gaussian integer factors $1 + i$ and $1 - i$, which shows that the ordinary prime 2 is not a Gaussian prime.

In fact, no ordinary integer of the form $a^2 + b^2$ is a Gaussian prime, because it has the Gaussian prime factorization

$$a^2 + b^2 = (a + bi)(a - bi).$$

This is actually good news, because the factorization of $a^2 + b^2$ reveals hidden properties of *sums of two squares,* a topic that has fascinated mathematicians since the discovery of the Pythagorean theorem. But first we need to understand *division* for the Gaussian integers, as we did for the ordinary integers.

Division of Gaussian Integers with Remainder

To divide a Gaussian integer A by a Gaussian integer B we do much the same as we did with ordinary integers: we see where A falls among all the multiples of B. The multiple QB nearest to A gives the quotient Q, and the difference $A - QB$ is the remainder. But is this remainder "smaller" than B? To answer this question we need to know what the multiples of B look like.

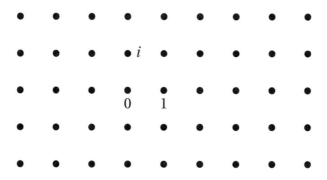

Figure 7.2: Gaussian integers.

The Gaussian integers form a square *grid* or *lattice* as shown in Figure 7.2. The squares have side length 1, because 1 and i both lie at distance 1 from O.

If B is any Gaussian integer, then the Gaussian integer $m + ni$ times B is m times B plus n times iB. Hence the multiples of B are the sums of the points B and iB, which lie at distance $|B|$ from O in perpendicular directions. Because of this, *the multiples of any Gaussian integer* $B \neq 0$ *form a lattice of squares of side* $|B|$. Figure 7.3 shows an example, the multiples of $2 + i$, shown as the black dots among the other Gaussian integers.

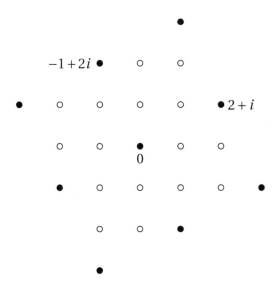

Figure 7.3: The lattice of multiples of $2 + i$.

Now let A be any Gaussian integer, and suppose we want to find its quotient and remainder on division by a Gaussian integer $B \neq 0$. The multiples of B form a square lattice and A falls within one of the squares. If QB is the corner of the square nearest to A we have $A = QB + R$, where $R = A - QB$, and $|R|$ is the hypotenuse of a right-angled triangle that fits in one *quarter* of the square (Figure 7.4).

Since the square has side length $|B|$, the sides of the right-angled triangle are both $\leq |B|/2$. Then the Pythagorean theorem gives

$$|R|^2 \leq \left(\frac{|B|}{2}\right)^2 + \left(\frac{|B|}{2}\right)^2 = \frac{|B|^2}{2}, \quad \text{and therefore} \quad |R| < |B|.$$

Putting these facts together, we have:

Division property of Gaussian integers. *For any Gaussian integers A and $B \neq 0$ there are Gaussian integers Q and R such that*

$$A = QB + R, \quad where \quad 0 \leq |R| < |B|.$$

(We do not claim that the Q and R are unique—in fact there are *four* nearest corners QB when A is at the center of the square—we need only the existence of *some* small remainder R.)

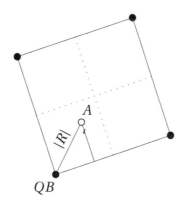

Figure 7.4: A and the nearest multiple of B.

Exercises

The question of "ordinary" division in the Gaussian integers—that is, division with zero remainder—boils down to the question of division in the natural numbers. This is due to the multiplicative property of absolute value, discussed in Section 2.5, which says that $|AB| = |A||B|$ for Gaussian integers A and B. It follows that $|AB|^2 = |A|^2|B|^2$, and of course all the numbers $|AB|^2$, $|A|^2$ and $|B|^2$ are natural numbers.

So, *a Gaussian integer C has a factorization into Gaussian integers A and B only if the natural number $|C|^2$ has a factorization into natural numbers $|A|^2$ and $|B|^2$.* This reduces the search for factors of a Gaussian integer to a search for ordinary factors of an ordinary integer.

7.5.1 If $C = 3 + 4i$, show that the prime factorization of $|C|^2$ is 5×5.

7.5.2 Explain why the only Gaussian integers A with $|A|^2 = 5$ are $\pm 2 \pm i$ and $\pm 1 \pm 2i$.

7.5.3 By choosing among the Gaussian integers found in the previous question, find A and B (not necessarily different) such that $C = AB$, with $|A|^2 = |B|^2 = 5$.

7.5.4 Find a factorization $5 + 7i = AB$ where $|A|^2 = 2$ and $|B|^2 = 37$.

7.6   Gaussian Primes

We know from Section 7.3 that the division property of ordinary integers paves the way for unique prime factorization, so we may expect the same with Gaussian integers. But first we need to define Gaussian primes, and confirm that Gaussian prime factorization exists. It generally does but, just as we exclude the number 1 from the ordinary primes, it is appropriate to exclude $\pm 1, \pm i$ (called *units*) from the Gaussian primes. This is done because of the way we measure "size" for complex numbers, and "size" is key in the discussion of primes.

An appropriate measure of size for Gaussian integers is the absolute value for complex numbers. We have already seen why in studying division with remainder. So we define a *Gaussian prime* to be a Gaussian integer, of absolute value > 1, which is not the product of Gaussian integers of smaller absolute value. Numbers of absolute value 1 are excluded—just as they were from the ordinary primes—to allow the possibility of unique prime factorization.

Prime Factorization: Existence

The existence of prime factorization is established as for ordinary integers in Section 7.1, with one modification. It helps to use the square of absolute value, called the *norm*, to measure size, rather than absolute value itself. This does not affect the concept of prime (since having greater norm is equivalent to having greater absolute value), but it has the advantage that the norm is an ordinary integer, so it clearly can decrease only finitely often.

The argument then goes as follows. If the Gaussian integer N is not itself a prime, then it is a product of smaller Gaussian integers of norm greater than 1, A and B say. If A or B is not itself a Gaussian prime, it too can be written as a product of smaller Gaussian integers of norm greater than 1, and so on. Since positive integers cannot decrease forever, in a finite number of steps we find N written as a product of Gaussian primes.   □

The norm also helps in the search for divisors of a Gaussian integer, because of its multiplicative property,

$$\text{norm}(B)\text{norm}(C) = \text{norm}(BC).$$

This is merely a recycling of the Diophantus identity from Section 2.5,

$$(a^2 + b^2)(c^2 + d^2) = (ac - bd)^2 + (bc + ad)^2,$$

with $B = a + bi$ and $C = c + di$. If we let $A = BC$, the multiplicative property says that *if B divides A, then norm(B) divides norm(A)*. Hence we can narrow the search for divisors of A by looking only at Gaussian integers whose norms divide $\text{norm}(A)$.

Examples

1. The Gaussian integer $1 + i$ has norm $1^2 + 1^2 = 2$, which is an ordinary prime, hence not a product of smaller ordinary integers. Hence $1 + i$ is not the product of Gaussian integers of smaller norm—it is a Gaussian prime. Similarly, $1 - i$ is a Gaussian prime, and hence $(1 + i)(1 - i)$ is a Gaussian prime factorization of 2.

2. The Gaussian integer 3 has norm 3^2, the ordinary divisors of which are 1, 3, and 3^2. But 3 is *not* the norm $a^2 + b^2$ of a Gaussian integer $a + bi$, since 3 is not a sum of squares of integers a and b. Hence 3 is not a product of Gaussian integers of smaller norm—it is a Gaussian prime.

3. The Gaussian integer $3 + 4i$ has norm $3^2 + 4^2 = 5^2$, and 5 is an ordinary prime. Hence any proper Gaussian factors of $3 + 4i$ have norm 5. Trying the obvious Gaussian integer with norm 5, namely $2 + i$, we find that $(2 + i)^2 = 2^2 + 4i + i^2 = 3 + 4i$. The factor $2 + i$ is a Gaussian prime because its norm is an ordinary prime.

Prime Factorization: Uniqueness

We have been poised to prove unique prime factorization for Gaussian integers ever since we proved their division property, so let us not hesitate any longer. It remains only to check that the procedure used in Section 7.4 for ordinary integers carries over to Gaussian integers.

- First, we need to show that, for any Gaussian integers A and B, there are Gaussian integers M and N such that

$$\gcd(A, B) = MA + NB.$$

  This follows from the division property: either via the Euclidean algorithm, as in Section 7.3, or by showing that all the numbers $MA + NB$ are multiples of their smallest member, as in Section 7.2. (where "smallest" now means smallest in absolute value).

- Next, we need the *Gaussian prime divisor property*: if a Gaussian prime P divides AB, then P divides A or P divides B. This follows as before, using $1 = \gcd(A, P) = MA + NP$ and multiplying the equation by B.

- Lastly, we would like to show that if $P_1 P_2 \cdots P_r$ and $Q_1 Q_2 \cdots Q_s$ are equal products of Gaussian primes, then they are products of the *same* primes. As before, we can use the prime divisor property to show that each P divides some Q. But this does not quite imply that $P = Q$, only that $P = \pm Q$ or $P = \pm iQ$. We say that P and Q are the *same up to unit factors*.

This means that prime factorization in the Gaussian integers is slightly "less unique" than in the positive integers. The exact statement is:

Unique prime factorization of Gaussian integers. *If $P_1 P_2 \cdots P_r$ and $Q_1 Q_2 \cdots Q_s$ are equal products of Gaussian primes, then the products are the same up to the order of factors and possible unit factors (± 1 and $\pm i$).*

Exercises

7.6.1 Find $|8 + 3i|^2$ and hence explain why $8 + 3i$ is a Gaussian prime.

7.6.2 Show that 61 is not a Gaussian prime by factorizing it into Gaussian integers of smaller absolute value.

7.6.3 Explain why $2 = (1 + i)(1 - i)$ is a Gaussian prime factorization.

7.6.4 Show that the factors A and B of $5 + 7i$ found in Exercise 7.5.4 are Gaussian primes.

7.6.5 By finding the ordinary prime factorization of $|11 + 3i|^2$, express $11 + 3i$ as a product of three Gaussian primes.

7.7 Rational Slopes and Rational Angles

A remarkably simple application of unique Gaussian prime factorization yields the result mentioned in Section 2.6: *the angle in a rational right-angled triangle is not a rational multiple of* π. (I learned of this proof from Jack Calcut.) Thus a sequence of rational right-angled triangles, such as those in Plimpton 322, can never produce a "scale" of equally-spaced angles between 0 and $\pi/2$, just as a sequence of rational frequency ratios cannot divide the octave into equal pitch intervals.

Suppose on the contrary that there is a triangle with integer sides a and b, integer hypotenuse c, and angle $2m\pi/n$ or some integers m and n. In other words, the point $\frac{a}{c} + \frac{bi}{c}$ lies on the unit circle at angle $2m\pi/n$ as in Figure 7.5.

It follows by the angle addition property of multiplication that

$$\left(\frac{a}{c} + \frac{bi}{c} \right)^n = 1,$$

and therefore

$$(a + bi)^n = c^n. \tag{1}$$

This equation violates unique Gaussian prime factorization, though it is not yet obvious why, since $a + bi$ and c are not necessarily Gaussian primes. But they have Gaussian prime factorizations

$$a + bi = A_1 A_2 \cdots A_r,$$
$$c = C_1 C_2 \cdots C_s, \tag{2}$$

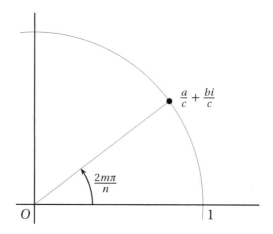

Figure 7.5: The hypothetical point.

which are unique up to the order of terms or unit factors. Also, $a + bi$ is not a multiple $\pm c$ or $\pm ic$ of c by a unit, so the list $A_1, A_2, \ldots, A_r$ differs from the list $C_1, C_2, \ldots, C_s$ by *more* than order or unit factors. (That is, the lists contain different primes, or different numbers of them, up to unit factors.)

Thus if we substitute (2) in (1) we get equal prime factorizations

$$(A_1 A_2 \cdots A_r)^n = (C_1 C_2 \cdots C_s)^n$$

that differ by more than order or unit factors. This contradicts unique Gaussian prime factorization. So our original assumption, that $\frac{a}{c} + \frac{bi}{c}$ has angle $2m\pi/n$ for some integers m and n, must be false. $\qquad\square$

Exercises

The result above shows that we cannot divide a right angle into equal parts by rational points on the unit circle. This explains why the "scale of angles" given in Figure 1.5 *necessarily* consists of unequal parts of the right angle.

Nevertheless there is a connection between Pythagorean triples and Gaussian integers—particularly the squares of Gaussian integers.

7.7.1 Show that $(2 + i)^2 = 3 + 4i$, and that $|3 + 4i| = 5$.

7.7.2 Show that $(3+2i)^2 = a+bi$, where $(a, b, |a+bi|)$ is a Pythagorean triple.

7.7.3 For any Gaussian integer $p+qi$, find the ordinary integers a and b such that $(p+qi)^2 = a+bi$, and show that $(a, b, |a+bi|)$ is a Pythagorean triple.

7.7.4 Find a Gaussian integer $p+qi$ whose square $a+bi$ gives the triple $(7, 24, 25)$ as in 7.7.3.

7.7.5 Investigate some of the triples in the Plimpton 322 table, shown in Section 1.2, in the light of these results. For example, which values of p and q give the triple $(120, 119, 169)$?

7.8   Unique Prime Factorization Lost

> It is greatly to be lamented that this virtue of the real numbers [the ordinary integers], to be decomposable into prime factors, always the same ones … does not also belong to the complex numbers [complex integers]; were this the case, the whole theory … could easily be brought to its conclusion. For this reason, the complex numbers we have been considering seem imperfect, and one may well ask whether one ought not to look for another kind which would preserve the analogy with the real numbers with respect to such a fundamental property.
>
> E. E. Kummer (1844), translated by André Weil in the introduction to Kummer's *Collected Papers*.

The Gaussian integers are well suited to explaining the angles in rational triangles, but other problems call for other "complex integers." The breakthrough example was Euler's solution of the equation $y^3 = x^2 + 2$ in his *Elements of Algebra* of 1770. The problem, already mentioned briefly in Section 7.5, is to show that the only positive integer solution of this equation is $x = 5$, $y = 3$. Diophantus mentioned this solution in Book VI, Problem 17 of his *Arithmetica*, and in 1657 Fermat claimed that it is the only one. Euler solved the problem by factorizing $x^2 + 2$ into "complex integers."

He begins by supposing that x and y are positive integers with

$$y^3 = x^2 + 2 = (x + \sqrt{-2})(x - \sqrt{-2}).$$

He then boldly assumes that, since $x^2 + 2$ is the cube y^3, its complex factors $x + \sqrt{-2}$ and $x - \sqrt{-2}$ are also cubes. In particular

$$x + \sqrt{-2} = (a + b\sqrt{-2})^3 \quad \text{for some ordinary integers } a \text{ and } b$$
$$= a^3 + 3a^2 b\sqrt{-2} + 3ab^2(-2) + b^3(-2)\sqrt{-2}$$
$$= a^3 - 6ab^2 + (3a^2 b - 2b^3)\sqrt{-2}.$$

Comparing the imaginary parts of both sides, one finds

$$1 = 3a^2 b - 2b^3 = b(3a^2 - 2b^2).$$

Now the only integer divisors of 1 are ± 1, so we must have

$$b = \pm 1, \quad 3a^2 - 2b^2 = \pm 1,$$

and therefore $a = \pm 1$, $b = \pm 1$. Substituting these in $x = a^3 - 6ab^2$ we find that only $a = -1$, $b = -1$ gives a positive solution, namely $x = 5$. The corresponding positive solution for y is obviously 3. $\qquad \square$

This miraculous proof succeeds because Euler was correct to assume that $x + \sqrt{-2}$ and $x - \sqrt{-2}$ are cubes in the universe of numbers $a + b\sqrt{-2}$, where a and b are ordinary integers. His reasons for this assumption are questionable—they also seem to apply to cases where such assumptions are false—but in this case complete justification is possible, thanks to *unique prime factorization for numbers $a + b\sqrt{-2}$.* This property holds here for much the same reasons that it holds in the Gaussian integers.

The *norm* of $a + b\sqrt{-2}$ is again the square of its absolute value, that is, the ordinary integer $a^2 + 2b^2$. *Primes* of the form $a + b\sqrt{-2}$ are numbers that are not products of numbers of smaller norm. Unique prime factorization follows as usual from the *division property*, which can be proved by the same method we used for the Gaussian integers. We look at the points of the form $a + b\sqrt{-2}$ in the plane of complex numbers and observe that they form a lattice of rectangles of width 1 and height $\sqrt{2}$. The division property follows from the easily proved geometric fact that the distance from any point in such a rectangle to the nearest corner is < 1.

From these examples one begins to see a general strategy of using complex integers to study ordinary integers of particular forms, to solve equations of particular forms, and so on. This is the "whole theory" Kummer was talking about in the quote at the beginning of this section. Kummer himself pursued the strategy as far as it would go in connection with *Fermat's last theorem,*[1] the claim that there are no ordinary integers x, y, z, and $n > 2$ satisfying the equation $x^n + y^n = z^n$.

Just before Kummer attacked this problem, a faulty general "proof" had been given by factorizing $x^n + y^n$ into "complex integers" of the form $x + y\zeta_n^k$, where $\zeta_n = \cos\frac{2\pi}{n} + i\sin\frac{2\pi}{n}$. Since the product of these factors is the nth power z^n, it is tempting to assume (as in Euler's solution of $y^3 = x^2 + 2$) that each factor is itself an nth power. However, this assumes unique prime factorization for the complex integers built from ζ_n, and Kummer discovered that *unique prime factorization fails for these integers* when $n \geq 23$.

It is remarkable that Kummer could see failure of unique prime factorization in such an intricate situation, but his reaction to this failure was even more remarkable: *he refused to accept it!* He believed that when there is no unique factorization into actual primes there must be a unique factorization into "ideal primes"—taking the word "ideal" from a geometric term for imaginary points. Before we try to believe in ideal primes ourselves, we should look at the simplest case that calls for them.

The Complex Integers $a + b\sqrt{-5}$

Complex integers of the form $a + b\sqrt{-5}$, for ordinary integers a and b, arise as complex factors of ordinary integers of the form $a^2 + 5b^2$. For example,

$$6 = 1^2 + 5 \times 1^2 = (1 + \sqrt{-5})(1 - \sqrt{-5}).$$

But 6 also equals 2×3, and it turns out that 2, 3, $1 + \sqrt{-5}$, and $1 - \sqrt{-5}$ are all primes—which spells trouble for unique prime factorization.

---

[1] Fermat's last theorem was eventually proved in the 1990s. The proof, by Andrew Wiles, was the culmination of a different line of attack through the theory of *elliptic curves*. There is no known proof using Kummer's strategy.

This is another case where it is appropriate to define the norm of a complex integer as the square of its absolute value, so $\text{norm}(a+b\sqrt{-5}) = a^2 + 5b^2$. As usual, a *prime* is not the product of complex integers of smaller norm, and B divides A in the complex integers only if $\text{norm}(B)$ divides $\text{norm}(A)$ in the ordinary integers. Thus $a + b\sqrt{-5}$ is prime if its norm cannot be split into smaller integer factors which are also norms. This gives a quick way to check that the numbers 2, 3, $1 + \sqrt{-5}$, and $1 - \sqrt{-5}$ are all primes:

- 2 has norm 2^2, which factorizes only as 2×2, and 2 is not a norm because $2 \neq a^2 + 5b^2$ for any ordinary integers a and b.

- 3 has norm 3^2, which factorizes only as 3×3, and 3 is not a norm because $3 \neq a^2 + 5b^2$ for any ordinary integers a and b.

- $1 + \sqrt{-5}$ has norm 6, which factorizes only as 2×3, and we have just established that 2 and 3 are not norms.

- $1 - \sqrt{-5}$ also has norm 6, so we can draw the same conclusion.

Thus the factorizations 2×3 and $(1 + \sqrt{-5})(1 - \sqrt{-5})$ of 6 are both prime factorizations, but the norm calculations we have just made tell us that *they fail the prime divisor property*. For example, 2 divides $6 = (1 + \sqrt{-5})(1 - \sqrt{-5})$ but 2 does *not* divide $1 + \sqrt{-5}$ or $1 - \sqrt{-5}$, because $\text{norm}(2) = 2^2$ does not divide $\text{norm}(1 + \sqrt{-5}) = \text{norm}(1 - \sqrt{-5}) = 6$.

The unacceptable behavior of these so-called primes convinced Kummer that they were not really prime, but compounds of hidden "ideal" primes, whose presence could be inferred from the behavior of the so-called primes. He compared the situation with the state of chemistry in his time, where an element "fluorine" was believed to exist, but was observed only in fluoride compounds. It was nevertheless possible to infer some properties of fluorine, and to predict the behavior of new fluorine compounds.

Kummer proceeded in this way with his ideal primes. He was able to gain the advantages of unique prime factorization without actually finding the ideal primes. And just as fluorine was eventually isolated, so (in a sense) were Kummer's "ideal primes," though not by Kummer himself.

Exercises

7.8.1 Find ordinary prime factorizations of the ordinary integers 10, 34, 65, and 85. Check in each case that the integer and all of its prime factors are sums of two squares.

The following exercises arrive at an explanation of this fact, which has to do with factorization into Gaussian integers.

For any Gaussian integer $A = m + in$ we call $\overline{A} = m - in$ the *conjugate* of A. (The conjugate was introduced previously, for arbitrary complex numbers, in Section 6.5.) Gaussian integers "behave the same" as their conjugates in the sense that

$$\overline{A} + \overline{B} = \overline{A + B} \quad \text{and} \quad \overline{A} \cdot \overline{B} = \overline{A \cdot B} \tag{*}$$

7.8.2 If $A = m + in$ and $B = p + iq$, calculate $\overline{A} + \overline{B}$ and $\overline{A} \cdot \overline{B}$. Hence prove (*).

7.8.3 Consider the factorization

$$m^2 + n^2 = (m + in)(m - in)$$

into conjugate Gaussian integers. If $m + in$ has Gaussian prime factors $A, B, C, \ldots$ explain why $m - in$ has Gaussian prime factors $\overline{A}, \overline{B}, \overline{C}, \ldots$.

7.8.4 If A is any Gaussian prime, explain why $A\overline{A}$ is an ordinary prime that is a sum of two squares.

7.8.5 Deduce from the previous two questions that any sum of two squares $m^2 + n^2$, for ordinary integers m and n, has a factorization into ordinary primes that are sums of two squares.

As we outlined above, unique prime factorization holds for the numbers $a + b\sqrt{-2}$, where a and b are ordinary integers, for much the same reason that it holds in the Gaussian integers. Here in more detail are the steps needed to justify this claim.

7.8.6 If $B = a + b\sqrt{-2}$ show that the multiples MB of B form a lattice of rectangles the same shape as the lattice of numbers M; namely, the height of each rectangle is $\sqrt{2}$ times its width.

7.8.7 In a rectangle of height $\sqrt{2}$ and width 1, show that any point is at distance < 1 from the nearest corner. (Hint: consider the center point.)

7.8.8 Explain how the previous exercise gives M and R such that

$$A = MB + R, \quad \text{where } |R| < |B|,$$

for any A and B of the form $a + b\sqrt{-2}$ (the division property).

Then, as we know, the division property implies a Euclidean algorithm, hence the prime divisor property, and hence unique prime factorization for numbers of the form $a + b\sqrt{-2}$.

7.9 Ideals—Unique Prime Factorization Regained

> But the more hopeless one feels about the prospects of later research on such numerical domains, the more one has to admire the steadfast efforts of Kummer, which were finally rewarded by a truly great and fruitful discovery.
>
> Richard Dedekind *Theory of Algebraic Integers*, p. 56.

If Kummer is right, the so-called primes 2 and $1 + \sqrt{-5}$ among the complex integers $a + b\sqrt{-5}$ behave badly because they are not genuine primes. They split into "ideal" factors, but how are such "ideal numbers" to be observed? Kummer's basic idea was that *a number is known by the set of its multiples*, so it is enough to describe the multiples of an ideal number. In 1871, Dedekind took this idea to its logical conclusion by working with the "sets of multiples" alone, which he called *ideals*. He found that ideals are simple to describe, easy to multiply, and that they have unique prime factorization for any reasonable domain of "complex integers"—a complete realization of Kummer's dream.

Dedekind's invention of ideals is one of the great success stories of mathematics, and not only because it restored unique prime factorization. Ideals turned out to be a fruitful concept in many parts of number theory, algebra, and geometry. They have been so overwhelmingly successful that many people now study them without having any idea of their original purpose—an absurd and unfortunate situation, since it trivializes a victory over the seemingly impossible. My objective here is to give a clear and convincing example of an ideal, which everyone should see *before* studying ideals in general.

Naturally, the place to look is among the complex integers $a+b\sqrt{-5}$, and at the numbers 2 and $1+\sqrt{-5}$ in particular. We suspect that 2 and $1+\sqrt{-5}$ split into ideal primes, so they may have an ideal factor in common. This prompts us to look for the *greatest common divisor* of 2 and $1+\sqrt{-5}$, $\gcd(2, 1+\sqrt{-5})$. We recall from Section 7.2 that the gcd of ordinary integers a and b is the least nonzero number among the integers $ma+nb$. In fact, we showed in Section 7.2 that the integers $ma+nb$ are precisely the integer multiples of $\gcd(a,b)$, so $\gcd(a,b)$ may be "known" by the set of integers of the form $ma+nb$, where m and n are ordinary integers.

It is then reasonable to expect that $\gcd(2, 1+\sqrt{-5})$ may be "known" by the set of numbers of the form $2M + (1+\sqrt{-5})N$, where M and N run through all the complex integers $a+b\sqrt{-5}$. Let us see what this set looks like.

First consider all the multiples $2M$ of 2, which form a lattice of rectangles of the same shape (but twice as large) as the lattice of all the complex integers $a+b\sqrt{-5}$. This set is shown in Figure 7.6.

Next, consider the set of multiples $(1+\sqrt{-5})N$ of $1+\sqrt{-5}$, which is shown in Figure 7.7.

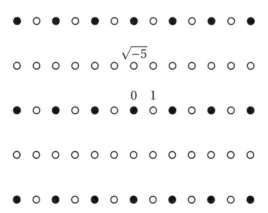

Figure 7.6: The multiples of 2.

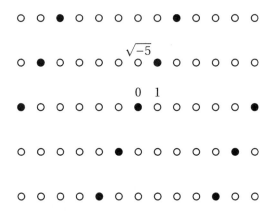

Figure 7.7: The multiples of $1 + \sqrt{-5}$.

These also form a lattice of the same shape, since multiplying the whole plane of complex numbers by the constant $1 + \sqrt{-5}$ magnifies all distances by a constant. (It is a little harder to spot the rectangles at first, since they are rotated by the imaginary component of $1 + \sqrt{-5}$.) The lattices $2M$ and $(1 + \sqrt{-5})N$ are examples of what are called *principal ideals*, a principal ideal in general being the set of multiples of some fixed number.

Finally, consider the sums $2M + (1 + \sqrt{-5})N$ of multiples of 2 and multiples of $1 + \sqrt{-5}$ (Figure 7.8). These sums form a lattice which is *not* rectangular, hence it is not the set of multiples of any number of the form $a + b\sqrt{-5}$.[2] In Kummer's language, its members are the multiples of the "ideal number" $\gcd(2, 1 + \sqrt{-5})$. In today's more prosaic language, the lattice is a *nonprincipal ideal*.

One finds lattices representing $\gcd(3, 1 + \sqrt{-5})$ and $\gcd(3, 1 - \sqrt{-5})$ in the same way. Each is a nonprincipal ideal.

---

[2] Kummer's discovery of the "multiples" of a nonexistent number, and his reaction to it, were described by Dedekind as follows (in §162 of his 1871 Supplement X to Dirichlet's *Vorlesungen über Zahlentheorie*):

> But it can very well be the case that no such number exists ... On meeting this phenomenon, Kummer had the fortunate idea to fake [fingieren] such a number and introduce it as an *ideal number*.

The word *fingieren* means to fake or fabricate, so the story of ideal numbers and their realization by ideals is a splendid illustration of the slogan "fake it until you make it!"

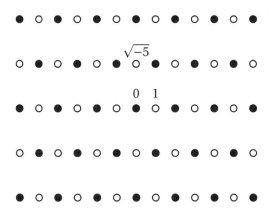

Figure 7.8: The multiples of 2 plus the multiples of $1 + \sqrt{-5}$.

Now any two ideals $\mathcal{I}$ and $\mathcal{J}$ have a natural *product* $\mathcal{I}\mathcal{J}$, consisting of all sums $A_1 B_1 + A_2 B_2 + \cdots + A_k B_k$ where the As are from $\mathcal{I}$ and the Bs are from $\mathcal{J}$. When $\mathcal{I}$ is the principal ideal (A) of all multiples of A and $\mathcal{J}$ is the principal ideal (B) of all multiples of B, then $\mathcal{I}\mathcal{J}$ is simply the principal ideal (AB) of all multiples of AB. Thus the product of principal ideals corresponds to the product of numbers, and we should regard the product of nonprincipal ideals as the product of "ideal" numbers. This is confirmed when we form products of the nonprincipal ideals $\gcd(2, 1+\sqrt{-5})$, $\gcd(3, 1+\sqrt{-5})$, and $\gcd(3, 1-\sqrt{-5})$: *we recover the numbers that these "ideal numbers" are supposed to divide.*

We skip the calculations, and just give the results:

$$\gcd(2, 1 + \sqrt{-5})\gcd(2, 1 + \sqrt{-5}) = (2)$$
$$\gcd(3, 1 + \sqrt{-5})\gcd(3, 1 - \sqrt{-5}) = (3)$$
$$\gcd(2, 1 + \sqrt{-5})\gcd(3, 1 + \sqrt{-5}) = (1 + \sqrt{-5})$$
$$\gcd(2, 1 + \sqrt{-5})\gcd(3, 1 - \sqrt{-5}) = (1 - \sqrt{-5}).$$

It follows that *the two factorizations of* 6, 2×3 *and* $(1 + \sqrt{-5})(1 - \sqrt{-5})$, *split into the same product of ideal numbers:*

$$\gcd(2, 1 + \sqrt{-5})\gcd(2, 1 + \sqrt{-5})\gcd(3, 1 + \sqrt{-5})\gcd(3, 1 - \sqrt{-5}).$$

Moreover, these ideal factors turn out to be prime, and ideal prime factorization turns out to be unique, so Kummer was right. The so-called primes $2, 3, 1 + \sqrt{-5}, 1 - \sqrt{-5}$ are really compounds of the ideal primes $\gcd(2, 1 + \sqrt{-5})$, $\gcd(2, 1 - \sqrt{-5})$, $\gcd(3, 1 + \sqrt{-5})$, and $\gcd(3, 1 - \sqrt{-5})$.

The ideal primes *do not exist* among the numbers $a + b\sqrt{-5}$ where a and b are ordinary integers. Yet, miraculously, their "sets of multiples" do and we can work with these sets almost as easily as we do with ordinary numbers. In 1877, Dedekind [13, p. 57] compared this with the situation of irrational numbers, such as $\sqrt{2}$. $\sqrt{2}$ does not exist among the rational numbers, but it can be modelled by a *set* of rationals (such as the set of decimal fractions we used in Section 1.5), and we can work with such sets almost as easily as with individual rationals.

Thus, in two important cases, we can realize "impossible" numbers by infinite sets of ordinary numbers. This discovery of Dedekind turned out to be very fruitful, and we pursue it further in Chapter 9.

Exercises

Since unique prime factorization does not hold for the complex integers $a + b\sqrt{-5}$ the division property cannot hold for them either. It is easy to see a geometric reason for this.

7.8.1 Explain why the multiples MB of one number $B = a + b\sqrt{-5}$ by all numbers M of this form make a lattice of rectangles of the same shape as the lattice of numbers M. Namely, it consists of rectangles whose height is $\sqrt{5}$ times their width.

7.8.2 Show that, in a rectangle of height $\sqrt{5}$ and width 1, the center point is at distance > 1 from each corner.

7.8.3 Deduce from the previous question that the point $1 + \sqrt{-5}$ is at distance > 2 from the nearest multiple of 2 among the numbers $a + b\sqrt{-5}$. Conclude that we cannot divide $1 + \sqrt{-5}$ by 2 and get a remainder of size < 2.

Chapter 8

Periodic Space

Preview

We begin with an impossibility in art rather than in mathematics: the self-feeding *Waterfall* of M. C. Escher (Figure 8.1). The impossible cycle of water can be blamed on a geometric object called the *tribar*, which does not exist in ordinary space because it contains a triangle with three right angles.

This poses an interesting mathematical challenge: *can the tribar exist in some other three-dimensional space?* Our goal is to show that it can, and the search for a tribar-friendly world leads us to a variety of *periodic spaces.* These are spaces that appear infinite from a viewpoint inside the space, but in which each object is seen infinitely often.

Consider the cylinder. If light rays travel along its geodesics, then creatures living in the cylinder will observe one and the same object over and over again. It will seem to them that they are living in an infinite plane, but certain strange objects in the cylinder will look quite ordinary to them. For example, a polygon with two sides and two right angles (which does exist in the cylinder!) appears as an infinite zigzag path in the plane.

Generalizing the cylinder, we land in a three-dimensional space called the *3-cylinder*. Unlike the ordinary cylinder, the 3-cylinder cannot be viewed from "outside." We must view it from "inside," as a periodic space, and *infer* the properties of objects that look periodic in the "inside" view. In this way we find that *the 3-cylinder contains a tribar.*

This exercise in visualizing a strange space leads us to wonder about the nature of spaces and periodicity in general. We have time for only a brief sketch of these ideas, but it brings us to the brink of *topology*, one of the most important fields in twentieth century mathematics.

Figure 8.1: Escher's *Waterfall*.

8.1 The Impossible Tribar

> It seems to me that we are all afflicted with an urge and
> possessed with a longing for the impossible. The reality
> around us, the three-dimensional world surrounding us, is
> too common, too dull, too ordinary for us. We hanker after
> the unnatural or supernatural, that which does not exist, a
> miracle.
>
> Maurits Escher in *Escher on Escher. Exploring the
> Infinite*, p. 135

M. C. Escher is the favorite artist of mathematicians, and his works
appear in many mathematics books. No doubt this is due to Escher's
frequent use of mathematical themes, but I believe that his longing for
the impossible also strikes a chord in us. When we look at a picture like
Waterfall we wish it could be true, because the situation it depicts has
such charm, ingenuity, and (somehow) logic. It seems as if Escher has
caught a glimpse of something from another world.

When one looks closely at *Waterfall* it becomes clear that it is based
on the impossible object shown in Figure 8.2. This object is known as
the *tribar* or *Penrose tribar*. Escher learned of it from the paper Impossible Objects by Lionel and Roger Penrose in the *British Journal of
Psychology* 49 (1958), pp. 31–33.

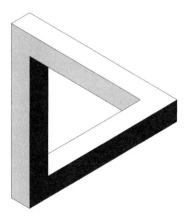

Figure 8.2: The tribar.

The Penroses (father and son) actually rediscovered the tribar. It has been known since the 1930s and is often used in popular art for paradoxical or whimsical effect. Variations of it go back centuries, to Piranesi and Bruegel, as one can see from the reproduction of Bruegel's *Magpie on the Gallows* in Figure 8.3. Is the crossbar of the gallows consistent with the position of its feet? One cannot be sure, and perhaps Bruegel wanted it that way.

Figure 8.3: Bruegel's *Magpie on the Gallows*.

With the tribar there is no such ambiguity. It cannot possibly exist in ordinary three-dimensional space, because there is no triangle with three right angles. However, *the tribar exists in other three-dimensional spaces*, and the aim of this chapter is to describe the simplest one. This is merely a project in recreational mathematics, since the tribar is not of great mathematical importance. Don't get any ideas about perpetual motion! However, it serves to introduce some important concepts of modern geometry and physics.

One concept that the tribar illustrates is the difference between "local" and "global." When viewed one small piece at a time, the tribar seems to be simply a square bar with right-angled corners. We might say that it is "locally consistent." But it is "globally inconsistent," at least with the ordinary three-dimensional space $\mathbb{R}^3$, where a triangle cannot have three right angles. The tribar needs an environment that is *locally* like $\mathbb{R}^3$, but globally different. To see why there is hope for this, recall Section 5.3, where we observed a distinction between local and global in two dimensions. The cylinder is locally the same as the plane, but globally different, and it differs in a way that makes it hospitable to certain paradoxical objects.

8.2 The Cylinder and the Plane

It is easy to construct figures on the cylinder that are impossible in the plane. We have already seen some in Section 5.3, such as closed straight lines and distinct lines that pass through the same two points. Figure 8.4 shows another paradox, similar in style to the tribar: a "two-sided polygon," or *2-gon*, with two right angles. A 2-gon does not exist in the plane, but it is clear from the figure that it exists in the cylinder.

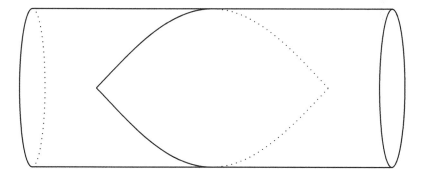

Figure 8.4: A right-angled 2-gon on the cylinder.

If we could roll up three-dimensional space the way we roll up the plane to form a cylinder, it might be possible to create a tribar. This is not physically possible, but we can see what to do if we look at the cylinder differently—not as a rolled-up strip of plane, but as a *periodic plane*. There is actually an ancient precedent for this view: it comes from Mesopotamia, the region "between the rivers" in what is now Iraq.

Among the treasures looted from the Iraqi National Museum in April 2003 were thousands of small stone cylinders, engraved with elegant designs featuring people, animals, and plants. These *cylinder seals*, as they are called, are among the finest works of art produced by the Mesopotamian civilization of 5,000 years ago. They were used to transfer a circular design (on the cylinder) to a periodic pattern on the plane. This was done by rolling the cylinder on soft clay, as indicated in Figure 8.5. Cylinder seals are not normally considered to be a mathematical achievement, but in a way they are, because they demonstrate that the cylinder is in some sense the same as a periodic plane.

Figure 8.5: Mesopotamian cylinder seal.

If we roll out the right-angled 2-gon design from Figure 8.4 it imprints a right-angled zigzag pattern on the plane as shown in Figure 8.6.

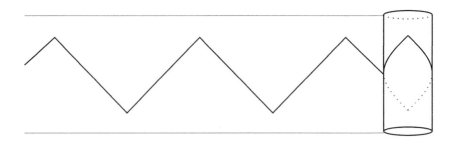

Figure 8.6: Zigzag pattern corresponding to the 2-gon.

A two-dimensional creature living in the cylinder would probably prefer to think of the cylinder as flat, because it *is* intrinsically flat, as we explained in Section 5.3. Such a creature would view the 2-gon as a zigzag path, but as it walked along the path the creature would experience an overpowering sense of déjà vu. No wonder, because each visit to the top vertex (say) would not only *seem* the same, it would actually *be* the same. The creature could confirm this by making a mark at the top vertex, and finding the same mark at the "next" top vertex. If it made more thorough experiments it would find that a mark at any point occurs again at a constant distance (for us, the circumference of the cylinder) in a certain fixed direction (for us, the direction perpendicular to the axis of the cylinder).

But if this creature had no sense of the third dimension it still would not be able to imagine the roundness of the cylinder. It would be easier for it to regard the cylinder as a periodic plane: a space in which each object is *repeated* over and over at regular spatial intervals. If light rays travelled along the geodesics of the cylinder, then the creature could actually see these repetitions, provided it looked in the right direction and there were no obstacles in the way. The view would be exactly the same as that of a creature in the periodic plane. It would not be as clear as our view of periodicity from *outside* the periodic plane (such as the view of the zigzag in Figure 8.6), since the whole field of view for a two-dimensional creature is only a line. To such a creature, periodicity would look something like Figure 8.7, which is a perspective view of a line divided into equal black and white segments.

Figure 8.7: Perspective view of a line.

Periodicity in three dimensions looks considerably more complex than this, as we already know from the view of flat space divided into equal cubes that was shown in Figure 5.1. However, we have to get used to it if we are to imagine the view from inside "cylindrical" three-dimensional spaces. We certainly cannot view them from outside!

Exercises

8.2.1 Show, using a suitable zigzag pattern, that there is a 2-gon on the cylinder whose sides meet at angle θ, for any θ between 0 and π.

8.2.2 Show, with the help of similar triangles in Figure 8.8 that the image of the point n on the x-axis is the point $\frac{n}{n+1}$ on the y-axis. (This is how Figure 8.7 was constructed.)

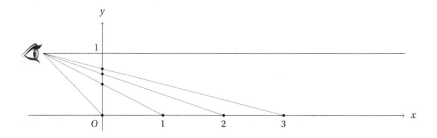

Figure 8.8: The eye's view of the line.

8.3 Where the Wild Things Are

The cylinder is a two-dimensional space that is like a circle in one direction and like a line in the perpendicular direction. A creature living in the surface of the cylinder would feel that it was living in a plane, except that the plane was periodic in one direction. This description is easily extended to a three-dimensional generalization of the cylinder, which we can call the *3-cylinder*. The 3-cylinder is a three-dimensional space that is like a circle in one direction, and like a plane in the directions perpendicular to the circle. A creature living in the 3-cylinder would feel that it was living in ordinary three-dimensional space, except that space was periodic in one direction.

What would this look like? The following picture by Magritte, Figure 8.9, is not quite the correct view (and no doubt he didn't have the 3-cylinder in mind), but it is a step in the right direction.

A man in the 3-cylinder, looking in the periodic direction, would see the back of his own head—just like the man looking in Magritte's creepy mirror. However, in the 3-cylinder the image of the back of his head would appear *infinitely* often, at equal intervals. If we simplify the head to a sphere, then the view in the periodic direction of the 3-cylinder could resemble Figure 8.10.

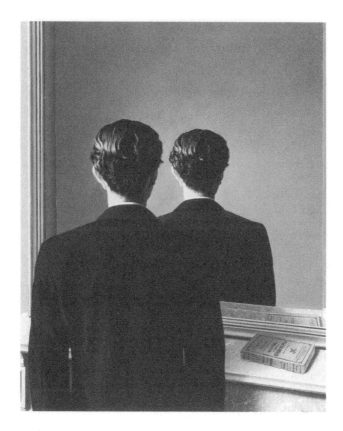

Figure 8.9: Magritte's *La reproduction interdite.*

Also appearing in this view is what looks like a periodic square bar with right-angled corners. But of course, since each sphere in the view is actually a new view of the *same* sphere, each corner of the bar next to the sphere is actually the *same* corner of a *closed* bar. Count the number of segments of the bar between repetitions and you will find there are three. The bar is actually a tribar! [1]

---

[1] Notice also, as the periodic square bar recedes into the distance, that the repeating sequence of three bars looks more and more like a sequence of tribars.

Figure 8.10: A tribar and a sphere in the 3-cylinder.

This explains, I hope, why the tribar really exists in the 3-cylinder. But in what sense does the 3-cylinder exist? Actual physical space is probably not a 3-cylinder, but it is probably not the "ordinary" three-dimensional space $\mathbb{R}^3$ either, so we can't expect an answer to this question from astronomy. However, there are several ways to realize the 3-cylinder mathematically. Each is like a realization of the ordinary cylinder, which we might call the *2-cylinder* to help the analogy along.

- Just as the 2-cylinder can be defined as an object in $\mathbb{R}^3$, consisting of all the points at distance 1 from the x-axis (say), the 3-cylinder can be defined as an object in $\mathbb{R}^4$ consisting of all points at distance 1 from the (x, y)-plane (say).

- Just as points on the 2-cylinder have coordinates (x,θ), where x is a real number and θ is an angle, points on the 3-cylinder have coordinates (x, y,θ), where x and y are real numbers and θ is an angle.

- Just as the 2-cylinder can be constructed by joining opposite sides of a strip of the plane bounded by two parallel lines, the 3-cylinder can be constructed by joining opposite sides of a slab of space bounded by two parallel planes.

The process of "joining" in the last item raises further questions about the meaning of mathematical constructions. Since these questions are crucial to many parts of modern mathematics, we shall discuss them at greater length in the next section.

Exercises

8.3.1 Given that the 3-cylinder shown in Figure 8.10 is formed by joining opposite sides of a slab, describe the position of the planes on opposite sides of the slab by reference to the spheres and bars in the figure.

8.3.1 Show that, for given x and y, the points $(x, y, \cos\theta, \sin\theta)$ in $\mathbb{R}^4$ all lie at distance 1 from $(x, y, 0, 0)$. (They form a circle of radius 1 on the 3-cylinder.)

8.4   Periodic Worlds

> The precise definition … uses the same trick that mathematicians always use when they want two things that are not equal to be equal.
>
> Michael Spivak, *A Comprehensive Introduction to Differential Geometry*, vol. 1, p. 13

"Joining" points that occupy different positions in space, or calling points the "same" when they are different, is done quite commonly in mathematics. How do mathematicians get away with it? Well, sometimes there is a clear physical realization of the joining process, as when an ordinary cylinder is made by joining opposite sides of a paper strip. Or sometimes two different motions lead to the same result, as when a rotation through 360° gives the same result as a rotation through 0°. Because of this we say that 0 and 360 are same angle (in degrees), or that 0 and 2π are the same angle (in radians).

However, it is generally confusing to call things equal when they are not, and the proper thing to do is to call them *equivalent*, and to consider the *class* of all things equivalent to a given object A. Then objects A and B are equivalent just in case their classes are equal, and *we are fully entitled to speak of equality between equivalence classes.*

To show the advantage of this way of thinking we use it to give a rigorous definition of the concept of angle. As just mentioned, when angles are measured in radians, the angle 0 is the "same" as the angle 2π. Thus we make 0 *equivalent* to 2π, and consequently to -2π, also to $\pm 4\pi, \pm 6\pi, \dots$ and so on. The "angle 0" is really the *equivalence class* $\{0, \pm 2\pi, \pm 4\pi, \dots\}$ of real numbers. In general, we define

$$\text{angle } \theta = \{\theta, \theta \pm 2\pi, \theta \pm 4\pi, \theta \pm 6\pi, \dots\}.$$

These numbers form an infinite series of points, at intervals of 2π along the number line. Figure 8.11 shows the equivalence class of 0 as a sequence of white dots, and the equivalence class of $\pi/2$ as a sequence of gray dots, together with the corresponding single white dot on the unit circle corresponding to the angle 0 and the single gray dot corresponding to the angle $\pi/2$.

Then equal angles, for example "angle $\pi/2$" and "angle $5\pi/2$," are *genuinely* equal equivalence classes. In this example they are both the class of gray points. Equivalence classes can also be *added*, in a way that makes sense of addition of angles. Namely, we let the class of θ plus the class of φ be the class of $\theta + \varphi$. For example, we want angle π plus angle $3\pi/2$ to be angle $\pi/2$, and it is, because the class of $\pi + 3\pi/2$ *is* the class of $\pi/2$.

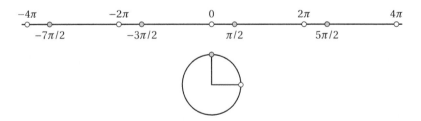

Figure 8.11: Angles as equivalence classes of numbers.

In describing angles as equivalence classes we are also describing points on the circle as equivalence classes, since each point on the circle corresponds to an angle. Figure 8.11 makes this clear: the white point on the circle corresponds to the white class on the line, the gray point on the circle corresponds to the gray class on the line, and so on. Thus the circle is a *periodic line*, in the same sense that the 2-cylinder is a periodic plane and the 3-cylinder is a periodic space.

Now that we know what a periodic line is, we can describe the 2-cylinder as a plane in which one axis is an ordinary line and the other is a periodic line. This accounts for the coordinates (x, θ): x is the coordinate on the ordinary line and θ is the coordinate on the periodic line, that is, an angle. Similarly, the 3-cylinder is a space in which two axes are ordinary lines and the third is a periodic line, hence we can give it the coordinates (x, y, θ).

Exercises

Two other important spaces whose "points" are equivalence classes are the *real projective line* and the *real projective plane*. These spaces are mathematical models of the line with a point at infinity, and the plane with a "horizon," that we used in the discussion of perspective drawing in Chapter 3.

The real projective line (Figure 8.12) can be viewed as the ordinary number line with an extra point "at infinity," which is "approached" as we move far away in either the positive or negative direction. The points of the projective line correspond to the *lines of sight* of an "eye" placed at point marked ∞ in the figure.

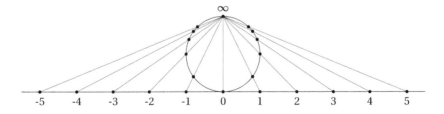

Figure 8.12: The projective line as a circle.

8.4.1 Explain why each finite point on the number line corresponds to a line of nonzero slope through the point ∞, while the point at infinity corresponds to the line through ∞ of slope zero.

8.4.2 If the lines through ∞ correspond to points on the circle as shown in Figure 8.12, explain why the point ∞ on the circle corresponds to the point "at infinity" of the number line.

The real projective plane similarly has "points" that correspond to lines through a point O in three-dimensional space (Figure 8.13)—the lines of sight of an eye viewing a plane—with the horizontal lines (appropriately) corresponding to points at infinity, and the plane of all horizontal lines through O corresponding to the horizon "line." More generally, any plane through O corresponds to a projective line in the projective plane. The "points" of the projective plane can be made to correspond to points on a sphere, but each such "point" corresponds to a *pair* of points on the sphere.

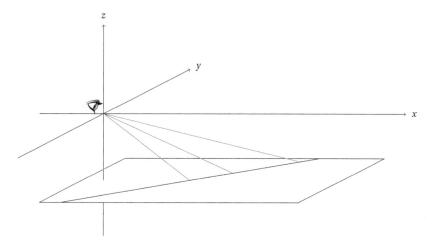

Figure 8.13: Viewing a plane from O.

8.4.3 Explain why each "point" of the real projective plane corresponds to a pair of *diametrically opposite* points on the unit sphere, (x, y, z) and $(-x, -y, -z)$, where x, y, z are numbers such that

$$x^2 + y^2 + z^2 = 1.$$

(Thus the real projective plane is a "periodic sphere," in which each point is equivalent to the diametrically opposite point.)

8.5   Periodicity and Topology

The freedom to replace ordinary lines by periodic lines enables us to construct surfaces or spaces that are periodic in more than one direction. For example, a plane with two periodic lines as axes is called a *torus* or *2-torus*. Just as we construct a cylinder from a strip of the plane by joining opposite sides, we can construct a 2-torus from a *square* by joining opposite sides as in Figure 8.14.

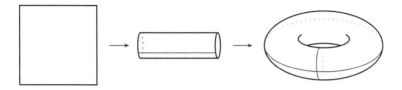

Figure 8.14: Construction of the 2-torus from a square.

The sides of the square are parallel to the axes and their length is the distance between equivalent points, so by joining the sides we are joining equivalent points, and there is exactly one point on the torus for each equivalence class of points in the periodic plane. The first stage, joining top to bottom, is exactly like the construction of a 2-cylinder. No distortion is required (it can be done with a strip of paper) and hence the resulting surface is still locally like the plane. However, the second stage *cannot* be done without distortion: joining the ends of the finite cylinder gives a surface with variable curvature. The resulting "bagel" surface is therefore not a geometrically faithful model of the 2-torus, unlike the periodic plane. Nevertheless, the bagel better exposes some *qualitative* properties of the 2-torus: the so-called *topological* properties. These include its finiteness, and the existence of closed curves that do not separate the surface, such as the two "joins."

Similarly, a space with three perpendicular periodic lines as axes is called a *3-torus*. Topologically, it is the result of joining opposite faces of a cube, though it is probably futile to try to visualize this process. It is better to visualize a space like ordinary space but periodic in three directions, which is easy to do with the help of Figure 8.15.

This figure, previously used in Chapter 5 as a picture of $\mathbb{R}^3$, can equally well be interpreted as the view from inside a 3-torus. Think of

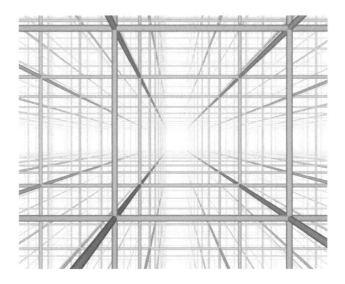

Figure 8.15: View from inside a 3-torus.

each cube as a repetition of the *same* 3-torus. If you were inside the 3-torus, you would see an image of yourself in every single cube—rather like what you would see from inside a cube with mirror walls, except that inside the 3-cube all your images face in the same direction and you can walk through the walls. Like the 2-torus, the 3-torus is a finite space, and it contains torus surfaces (each wall of the cube is one) that do not separate it: they have no "inside" or "outside." Also like the 2-torus, the 3-torus is a perfectly smooth space. The lines are simply "marks" drawn on it, not an intrinsic part of the space.

These surfaces and spaces are some of the simplest examples of *manifolds*: spaces that are locally similar to ordinary space but globally "wrapped around." Their key properties are topological, so a whole discipline of *topology* has been developed to study them. Just recently, topologists have begun to collaborate with astronomers to try to decide the question: which manifold is the physical universe? It is generally supposed that the universe is finite, and hence it cannot be like $\mathbb{R}^3$ or the 3-cylinder. It could be topologically the same as a 3-sphere, or a 3-torus, because both of these are finite. But there are infinitely many other options, involving more complicated periodicity. The picture used to depict hyperbolic space in Chapter 5, which we reproduce here as Figure 8.16, represents one possibility.

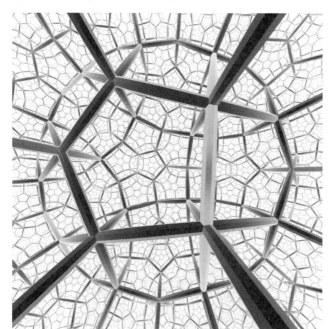

Figure 8.16: Dodecahedral space.

The picture shows a *non-Euclidean* periodicity, and it can be interpreted as a view from inside *dodecahedral space*, a manifold obtained by joining opposite faces of a dodecahedron. Each dodecahedral cell in the picture is a repetition of the same dodecahedral space.

With the more powerful telescopes, we may one day *observe periodicity in the universe*, and hence determine its topological nature. For more on this mind-bending possibility, and for a gentle introduction to three-dimensional manifolds, see the book *The Shape of Space* by Jeffrey Weeks. I also recommend Weeks' free software for viewing three-dimensional spaces, available at the web site

www.geometrygames.org/CurvedSpaces/.

Among the spaces you can view are the 3-torus and dodecahedral space, and some spaces for which the inside view is like a periodic 3-sphere.

Physicists are also interested in manifolds with more than three dimensions. It has been speculated that space actually has several more dimensions, but that these dimensions are periodic with period too small for us to observe directly. For more on these speculations, whose mathematical context is called *string theory*, see the book *The Elegant Universe* by Brian Greene.

Exercises

Figure 8.17 shows a square *PQRS* which, when opposite sides are joined as in Figure 8.14, becomes a torus. Also, the pieces labelled 1, 2, 3, 4, 5 become five squares on the torus, arranged in a way which is impossible for squares in the plane.

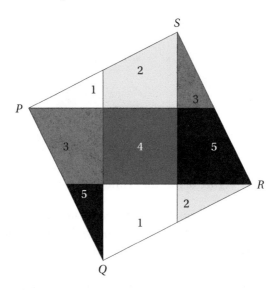

Figure 8.17: Torus divided into five squares.

8.5.1 Explain why the two pieces labelled 1 represent a single square on the torus, and similarly for the pieces labelled 2, 3, and 4.

8.5.2 Check that each one of the five squares on the torus shares a side with the other four.

8.5.3 Work out the numbers V and E of vertices and edges in this subdivision of the torus surface, and show that $V - E + F = 0$ (where, of course, $F = 5$ is the number of squares).

If this arrangement of squares is colored according to the usual rule for map coloring—that adjacent regions must have different colors—then five colors are needed. This cannot happen in the plane or sphere, where every map can be colored by at most four colors. This strange map of squares on the torus leads us to ask whether there are strange maps on the real projective plane, whose points, as we saw in Exercise 8.4.3, can be regarded as pairs of diametrically opposite points on the sphere.

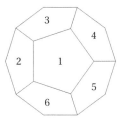

Figure 8.18: The front faces of the dodecahedron.

8.5.4 Show, by reference to Figure 8.18, that the projective plane can be tiled with six pentagons, three of which meet at each vertex. (Hint: Explain why each of the six front faces shown is the "same" as a face on the back.)

8.5.5 Supposing that each face on the back receives the same label as its opposite face on the front, show that the face labelled 6 shares edges with faces labelled 1, 2, 3, 4, 5.

8.5.6 Deduce that each of the six pentagons on the projective plane shares an edge with the other five, and hence that a coloring of the corresponding map requires six colors.

8.5.7 Show that the $F = 6$ pentagons between them share $E = 15$ edges and $V = 10$ vertices, so that $V - E + F = 1$.

We have now found, in particular cases, that the quantity $V - E + F$ equals 2 for the sphere, 1 for the real projective plane, and 0 for the torus. This number is in fact a characteristic number for these surfaces, known as the *Euler characteristic*. It retains its value for arbitrary arrangements of polygons.

8.6   A Brief History of Periodicity

Periodicity in time and in circular motion have always been with us, in the apparent rotation of the sun, moon, and stars in the sky. As a point moves with constant speed around the unit circle, its horizontal and vertical distances from the center vary periodically, and the functions describing this variation with angle θ are the *circular functions* cosine and sine (Figure 8.19).

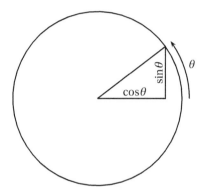

Figure 8.19: The cosine and sine as functions of angle.

As functions of time t, the horizontal (cosine) and vertical (sine) distances y of the moving point vary periodically between -1 and 1 as shown in Figure 8.20, with the cosine in gray and the sine in black. The two curves have the same shape—as one would expect from the symmetry of the circle—called a *sine wave*.

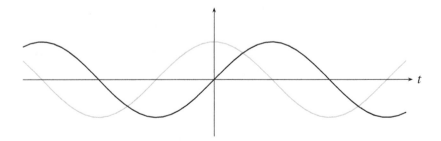

Figure 8.20: The cosine and sine as functions of time.

The sine wave is virtually the whole story of one-dimensional periodicity. In 1713 the English mathematician Brook Taylor discovered that the sine wave (scaled down in the y direction) is the shape of a string in its simple modes of vibration (shown in Figure 1.2 of this book). In 1753 the Swiss mathematical physicist Daniel Bernoulli correctly guessed, from the experience of hearing sounds resolve into simple tones, that *any* mode of vibration is a sum of simple modes. It follows that any reasonable periodic function on the line is a (possibly infinite) sum of sine waves. In 1822 the French mathematician Joseph Fourier refined Bernoulli's guess into a branch of mathematics now known as *Fourier analysis.* In a nutshell, Fourier analysis says that what you hear is what you get: one-dimensional periodicity reduces to sine waves, and hence ultimately to properties of the circle.

Periodicity in two dimensions, or on the complex numbers, is a comparatively recent discovery. It emerged from early calculus, and in particular from attempts to find the arc length of curves. As we saw in Chapter 4, this is a formidable problem even for the circle, though it can be solved with the help of integration and infinite series. Problems of greater difficulty begin with the arc length of the ellipse, and they lead to what are called *elliptic integrals* and *elliptic functions.*

An example of an elliptic integral is the arc length l of an ellipse from a given starting point, and an example of an elliptic function is the height of a point on the ellipse as a function $f(l)$ of arc length l (Figure 8.21).

Since the ellipse is a closed curve it has a total length, λ say, and therefore $f(l + \lambda) = f(l)$. The elliptic function f is *periodic*, with "period" λ, just as the sine function is periodic with period 2π. However, as Gauss discovered in 1797, elliptic functions are even more interesting than this: they have a second, *complex* period. This discovery completely changed the face of calculus, by showing that some functions should be viewed as functions on the *plane* of complex numbers. And just as periodic functions on the line can be regarded as functions on a periodic line—that is, on the circle—elliptic functions can be regarded as functions on a *doubly periodic plane*—that is, on a 2-torus. In the 1850s, Gauss's student Bernhard Riemann showed that the torus can be made the *starting point* of the theory of elliptic functions, so that their double periodicity becomes as natural as the single periodicity

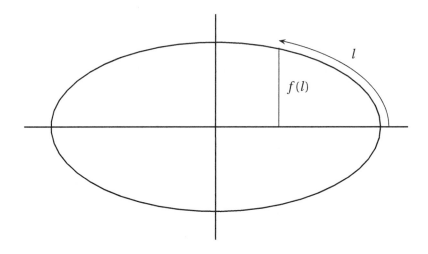

Figure 8.21: Example of an elliptic function.

of the circular functions. (It would be better to call elliptic functions "toral functions," but we seem to be stuck with the name "elliptic functions" for historical reasons. The same applies to the "elliptic curves" mentioned in Section 7.8.)

Double periodicity is more interesting than single periodicity, because it is more varied. There is really only one periodic line, since all circles are the same up to a scale factor. However, there are infinitely many doubly periodic planes, even if we ignore scale. This is because the angle between the two periodic axes can vary, and so can the ratio of period lengths. The general picture of a doubly periodic plane is given by a *lattice* in the plane of complex numbers: a set of points of the form $mA + nB$, where A and B are nonzero complex numbers in different directions from O, and m and n run through all the integers. A and B are said to *generate* the lattice because it consists of all their sums and differences.

Figure 8.22 shows an example. The lattice consists of the black dots, and these are all the points equivalent to O. The equivalence class of any point P is "P plus the lattice"—the points of the form $P + mA + nB$ (gray dots). Each equivalence class corresponds to a single point on the 2-torus.

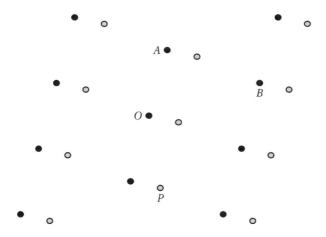

Figure 8.22: The lattice generated by A and B.

Exercises

Figure 8.23 shows a plane divided into squares, colored in a doubly-periodic manner.

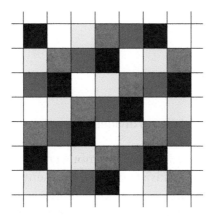

Figure 8.23: Periodic plane for the torus divided into five squares.

8.6.1 Explain, by drawing a square $PQRS$ suitably placed on this periodic plane, that it represents the torus divided into five mutually-adjacent squares found in the exercises to Section 8.5.

8.6.2 Notice again (and perhaps more easily) how each colored region is adjacent to the four other colors.

There is another interesting periodic plane defining a torus divided into seven mutually adjacent hexagons. Figure 8.24 shows a rhombus *PQRS*, from which the torus is constructed by "joining opposite edges," superimposed on the hexagon tessellation of the plane.

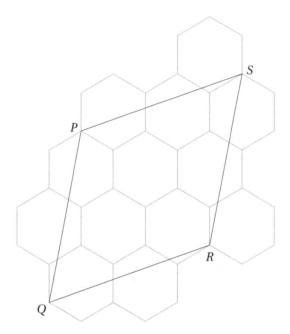

Figure 8.24: Torus divided into seven hexagons.

Notice that each side of the rhombus crosses three hexagons, so each of its four edges consists of three segments. Notice that corresponding segments on opposite sides—for example, the top segment of *PQ* and the top segment of *SR*—cut a hexagon into identical pieces. Hence the hexagon piece next to the top segment of *PQ* and *inside* the rhombus equals the hexagon piece next to the top segment of *SR* and *outside* the rhombus.

8.6.3 Deduce that the four edges *PQ*, *QR*, *RS*, and *SP* are equal (which is why we say *PQRS* is a rhombus).

8.6.4 By comparing hexagon pieces inside and outside the rhombus, show that the pieces inside the rhombus (including two complete hexagons) amount to exactly seven hexagons.

8.6.5 Conclude that the torus formed by joining opposite edges of the rhombus $PQRS$ is divided into seven hexagons. Check that each of these hexagons shares an edge with the six others. (Thus a coloring of the corresponding map requires seven colors.)

8.7 Non-Euclidean Periodicity

How many "essentially different" doubly periodic planes are there? The answer is: as many as there are *points* in the plane. We consider doubly periodic planes to be different if their lattices have a different shape, and we can describe each shape by a complex number, namely A/B. In fact, the number A/B captures the shape of the parallelogram with adjacent sides OA and OB (Figure 8.25). Its absolute value $|A|/|B|$ is the ratio of the side lengths, its angle $\alpha - \beta$ is the angle between the sides, and clearly the side length ratio and the angle determine the shape. (If you need to brush up on division of complex numbers, look back to the discussion of multiplication of complex numbers in Section 2.6.)

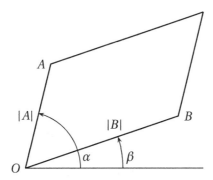

Figure 8.25: Why A/B captures the parallelogram shape.

The shape of the lattice of points $mA + nB$ can therefore be represented by the complex number A/B. It is not hard to see that any nonzero complex number represents a lattice shape, so in some sense there is a whole *plane of lattice shapes.* Even more interesting: *the plane of lattice shapes is a periodic plane, because different numbers represent the same lattice.* This is obvious when you think about it. For example, the lattice generated by A and B is also generated by $A + B$ and B, and hence the lattice represented by the number $C = A/B$ is also represented by the number $(A+B)/B = (A/B)+1 = C+1$. Likewise, the lattice generated by A and B is also generated by $-B$ and A, hence the lattice represented by the number $C = A/B$ is also represented by the number $-B/A = -1/C$. Thus, for any number $C \neq 0$ in the complex plane, the numbers $C+1$ and $-1/C$ are *equivalent* to C in the sense that they represent the same lattice shape.

It can be shown that all the numbers equivalent to C can be reached by these two operations: adding 1 and taking the negative reciprocal. Adding 1 of course is just the usual type of periodicity, but when this operation is intertwined with taking the negative reciprocal, the result is the complicated periodicity shown in Figure 8.26.

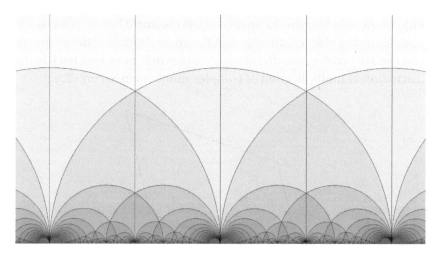

Figure 8.26: Periodicity of the plane of lattice shapes.

Only the upper half plane is shown, because each lattice shape can be represented by a number in the upper half plane. The pale blue triangle in the middle, which has one vertex at 0 and the other two on the lines $x = 1/2$ and $x = -1/2$, includes one number representative of each lattice shape. The other regions are obtained from it by repeatedly adding 1 or taking the negative reciprocal. For example, the infinite gray region directly above the middle triangle consists of the negative reciprocals of all the points inside the middle triangle. Notice also that the edges of the triangles are non-Euclidean lines in the sense of Section 5.7.

This pattern of periodicity was first glimpsed by Gauss around 1800. He was interested in lattice shapes in connection with elliptic functions, but also in connection with number theory (recall the lattices that came up in connection with complex integers in Chapter 7). However, he did not publish these ideas, and his results were later rediscovered by others. The double periodicity of elliptic functions was first published by the Norwegian mathematician Niels Henrik Abel and the German mathematician Carl Gustav Jacobi in the 1820s. The periodicity of the plane of lattice shapes reemerged from the study of the *elliptic modular function*, a function which is essentially a function of lattice shapes, introduced by Jacobi in the 1830s.

Other functions with similar periodicity were encountered in the 1870s, and in 1880 the French mathematician Henri Poincaré realized what they had in common: the periodicity of *regular tilings of the hyperbolic plane*. This was the first discovery of hyperbolic geometry in pre-existing mathematics. It led to a feverish exploration of the world of non-Euclidean periodicity by Poincaré and his German colleague Felix Klein. The very idea of a "periodic plane" (though not those words) dates from this period. Klein saw that the complexity of non-Euclidean tiling can be managed better by treating each class of equivalent points in the plane as a single point on a surface. He constructed this surface by joining equivalent points on the boundary of a tile—a polygon in the hyperbolic plane that includes a representative of each equivalence class—just as we constructed the torus from a square piece of the doubly periodic Euclidean plane in Section 8.5.

The simplest example of a surface whose "points" are equivalence classes in the hyperbolic plane is the *pseudosphere* we know from Sections 5.6 and 5.7. It is a hyperbolic analogue of the cylinder, obtained by joining equivalent points on opposite sides of a wedge of the hyperbolic plane bounded by asymptotic lines. (In Figure 8.26 the vertical lines are asymptotic hyperbolic lines. They become closer and closer to each other as their height increases.)

The surfaces arising from periodicity in the hyperbolic plane are like the cylinder or torus, but generally more complicated; they can have any number of "holes." When the idea is extended to spaces of three or more dimensions, as was done by Poincaré in 1895, the complexity of the resulting objects is even greater. A whole new branch of mathematics is required to deal with the qualitative geometry of periodicity, and *algebraic topology*, as it is now known, was created for this purpose by Poincaré. The topology of three-dimensional space is still not thoroughly understood, and one of Poincaré's questions about it—a problem about the 3-sphere called the *Poincaré conjecture*—became one of the most famous problems in mathematics.[2]

Exercises

The pseudosphere is the trumpet-shaped surface shown in Figures 5.21 and 5.24. Since this surface exists in ordinary space, we can easily visualize it and guess some of its properties. However, the behavior of the "lines" on the pseudosphere—the curves of shortest distance, also known as geodesics—is not so easy to guess. For example, are there geodesics that wind infinitely often around the pseudosphere, like the helices on the cylinder (Figure 5.11)? This question is answered by the half-plane model of the non-Euclidean plane, shown in Figures 5.27 and 8.26.

In the half-plane model, the pseudosphere is represented by a vertical strip, between two vertical lines and above a horizontal line that lies above the boundary of the half-plane (Figure 8.27). The intuition is that the pseudosphere is made by joining the left side of the strip to the right, to form a sort of half cylinder. The lower edge of the strip thereby becomes the circular boundary of the pseudosphere, and the cylinder tapers towards the top, because objects in the half-plane model appear larger as one moves higher in the vertical direction (look again at Figure 5.27).

Geodesics on the pseudosphere correspond, in exceptional cases, to vertical lines in the half-plane, such as the two edges of the strip,

---

[2]When the first edition of this book was written, a proof of the Poincaré conjecture had recently been announced by the Russian mathematician Grigory Perelman. Perelman's proof, which in fact covers a stronger claim called the *geometrization conjecture*, has now been checked and found to be correct.

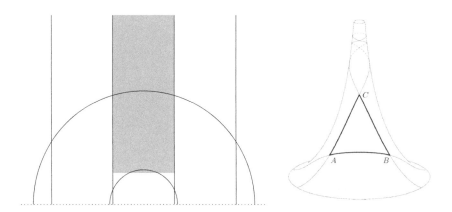

Figure 8.27: Geodesics on the pseudosphere.

which become a single geodesic on the pseudosphere. But generally
they correspond to semicircles with their centers on the boundary of
the half-plane, such as all the semicircles in Figure 8.26.

8.7.1 Deduce that the only geodesics that wind around the pseudosphere are those corresponding to semicircles in the half-plane,
such as the two shown in Figure 8.27. (The picture on the right,
which shows how the two geodesics look on the standard pseudosphere, is from Ratcliffe's *Foundations of Hyperbolic Manifolds*, p. 6.)

8.7.2 Identify, in the half-plane, the geodesic that goes through A, then
B, and the one that goes through A, then C, then B.

8.7.3 Deduce, further, that for any positive integer n there are geodesics
that wind n times around the pseudosphere.

8.7.4 Show that there is no geodesic that winds infinitely often around
the pseudosphere.

Chapter 9

The Infinite

Preview

The lesson of previous chapters seems to be that all roads in mathematics lead to infinity. At any rate, most of the attempts to do the impossible have called upon infinity in one way or another: not necessarily the infinitely large, not necessarily the infinitely small, but certainly the infinitely many.

We can do without infinite numbers, but not without infinite *sets*, especially the set $\mathbb{N} = \{1, 2, 3, 4, 5, \ldots\}$ of natural numbers, and the set $\mathbb{R}$ of real numbers, from which we construct the plane $\mathbb{R}^2$, Euclidean space $\mathbb{R}^3$, and various curves and surfaces that lie within them.

We first met $\mathbb{R}$ as the line, but we have also met important points on it, such as the integer and rational points, and the irrational points $\sqrt{2}$ and π. In Chapter 1 we saw how $\sqrt{2}$ could be understood as a *gap* in the infinite set of rational numbers. Here we attempt to broaden this understanding to cover all irrational numbers at once, and hence understand the line as a set of points.

This confronts us with a seemingly impossible situation, since $\mathbb{R}$ cannot be grasped one element at a time, but only as a whole.

To prepare to meet this difficulty we first examine the so-called *countable* sets, such as $\mathbb{N}$, which *can* be grasped one element at a time. The countable sets also include the set of rationals, but *not* the set of irrationals, and hence not the set $\mathbb{R}$ either.

We explain why $\mathbb{R}$ is not countable, showing in fact that each countable set occupies "almost none" of the line. We also point out some consequences of this situation, such as the existence of numbers that are not algebraic. Such numbers are very hard to find individually, but they include "almost all" real numbers.

9.1 Finite and Infinite

> The infinite we shall do right away. The finite may take a little longer.
>
> Stan Ulam, quoted in D. MacHale, *Comic Sections* (Dublin 1993)

As far as we know, the universe is finite. According to current cosmological beliefs, the universe began a finite number of years ago, from zero size, and has since expanded at a bounded rate. It therefore has a finite age, a finite size, and contains a finite number of particles.

Human artifacts, such as mathematical proofs, are also finite. So a proof that infinity exists can itself live in a finite universe, where what it "proves" is false. Hence the existence of infinity cannot be proved. And yet ... it is such a *convenient* assumption.

It should be clear from the preceding chapters of this book that infinity is the first refuge of the mathematician. Some of our favorite numbers, such as $\sqrt{2}$ and π, can be described only by infinite processes. Infinity is where parallel lines meet. An "ideal number" is represented by an infinite set of actual numbers, and an illusory finite object—the tribar—is represented by an actual infinite object, a periodic bar. And as Stan Ulam suggested in the remark quoted above, problems about the infinite are often *easier* than problems about the finite. For example, we found in Section 4.8 that the sum of the infinite series $1 - \frac{1}{3} + \frac{1}{5} - \frac{1}{7} + \cdots$ is $\pi/4$. Would you care to find the sum of the finite series $1 - \frac{1}{3} + \frac{1}{5} - \frac{1}{7} + \cdots + \frac{1}{1000001}$?

Thus "yearning for the impossible" is in many cases a yearning for the infinite, and to understand how mathematical dreams are realized one needs to understand a little about infinity.

I deliberately say "a little," because I want to keep this chapter as short as possible. The most remarkable thing about infinity is that we seem able to capture it in a finite number of words, so the less said the better if I am to make my point. Here are the kind of words I mean:

> *Consider the sequence* $1, 2, 3, 4, 5, \ldots$, *in which every number has a successor.*

Most of us are happy to consider this endless sequence, because it is *harder* to believe in the alternative—that the sequence ends in a number with no successor. It is like the situation posed in Lucretius' argument for infinite space (Section 5.1): if you think there is an end, what happens if you go there and hurl a javelin? (Or, in this case, go to the last number and add one?) There is no escaping infinity through periodicity either, since adding one always produces a number which is larger, and hence new. Admittedly, the assumption that each number has a successor cannot be proved; but the opposite assumption cannot be proved either, and it is far less plausible or useful.

The more one thinks about it, the clearer it becomes that the starting point for all mathematics is the infinite sequence $1, 2, 3, 4, 5, \ldots$ of *natural numbers.* The sequence is infinite in the sense of being produced by a process without end—start with 1 and keep adding 1. Until the nineteenth century it was not thought necessary to consider the *set* $\mathbb{N} = \{1, 2, 3, 4, 5, \ldots\}$ of all natural numbers as a completed whole. In fact, such thinking was deplored by many mathematicians and philosophers.

Exercises

One case in which the infinite *can* be approached through the finite is the infinite geometric series, as we saw in Section 4.3. Here, in more detail, are the steps in the argument.

9.1.1 Let

$$S_n = a + ar + ar^2 + ar^3 + \cdots + ar^n$$

and write down an expression for rS_n.

9.1.2 By calculating $S_n - rS_n$, show that

$$S_n = \frac{a - ar^{n+1}}{1 - r} \quad \text{when } r \neq 1.$$

What is the value of S_n when $r = 1$?

9.1.3 Since r^{n+1} approaches 0 as n increases when $|r| < 1$, conclude that S_n approaches $\frac{a}{1-r}$ as n increases.

The value $\frac{a}{1-r}$ being approached by $a + ar + ar^2 + ar^3 + \cdots + ar^n$ as n increases is what we mean by the *infinite sum* $a + ar + ar^2 + ar^3 + \cdots$.

9.2   Potential and Actual Infinity

> As to your proof, I must protest most vehemently against
> your use of the infinite as something consummated, as this
> is never permitted in mathematics. The infinite is but a
> figure of speech: an abridged form for the statement that
> limits exist which certain ratios may approach as closely
> as we desire, while other magnitudes may be permitted to
> grow beyond all bounds.
>
> Gauss, letter to Schumacher, 12 July 1831.

The sequence $1, 2, 3, 4, 5, \ldots$ of natural numbers is the model for all infinite sequences: it has an *initial term* (1) and a *generating process* (add 1) for going from each term to the next. Other examples are

$$1, \quad \frac{1}{2}, \quad \frac{1}{4}, \quad \frac{1}{8}, \quad \frac{1}{16}, \quad \frac{1}{32}, \quad \cdots$$

(initial term 1; each term is half the one before),

$$2, \quad 3, \quad 5, \quad 7, \quad 11, \quad 13, \quad 17, \quad \ldots$$

(initial term 2; each term is the next prime after the one before),

$$1, \quad 1 - \frac{1}{3}, \quad 1 - \frac{1}{3} + \frac{1}{5}, \quad 1 - \frac{1}{3} + \frac{1}{5} - \frac{1}{7}, \quad 1 - \frac{1}{3} + \frac{1}{5} - \frac{1}{7} + \frac{1}{9}, \quad \cdots$$

(initial term 1; each term adds the next term in the infinite series for $\pi/4$).

The notation for a general infinite sequence, $x_1, x_2, x_3, x_4, x_5, \ldots$, is clearly modelled on the sequence of natural numbers. For each natural number n there is an "nth term" x_n, so listing the terms in an infinite sequence is a process parallel to *counting* "$1, 2, 3, 4, 5, \ldots$." In fact, any set whose members can be arranged in an infinite sequence is called *countable* for just this reason. The set is "counted" by specifying a first member, second member, third member, ..., and arranging things so that each member occurs at some natural number stage (and hence each member has only finitely many predecessors).

Counting a set in this way arranges its members in a certain order, though the counting order may not be the natural order of the set. For example, take the set of rationals between 0 and 1. In the natural ordering of numbers this set has no first member, and between any two members there are infinitely many others. Nevertheless, it is possible to list the rationals between 0 and 1 by taking 1/2 first, then the fractions 1/3, 2/3 with denominator 3, then the fractions 1/4, 3/4 with denominator 4, and so on. The terms with denominator ≤ 6 are

$$\frac{1}{2}, \quad \frac{1}{3}, \frac{2}{3}, \quad \frac{1}{4}, \frac{3}{4}, \quad \frac{1}{5}, \frac{2}{5}, \frac{3}{5}, \frac{4}{5}, \quad \frac{1}{6}, \frac{5}{6}, \quad \cdots$$

and each rational number m/n between 0 and 1 occurs at some finite stage, since there are only finitely many rational numbers with denominator $\leq n$. It is also not hard to modify this idea to obtain a list of *all* rational numbers (hint: list according to the *sum* of absolute values of numerator and denominator). More surprisingly, perhaps, one can list all *algebraic* numbers, as we shall see in Section 9.5. Thus it begins to look as though all infinite sets are countable and hence fairly easy to grasp.

A countable set is infinite, but only "potentially" so, because each member appears at some finite stage. One need not try to grasp all members of the set simultaneously—just the process for producing them. From ancient times until the nineteenth century this was considered the only acceptable use of infinity in mathematics. In fact, until 1874 it may have seemed the only use necessary, because until that year only countable sets were known.

However, Gauss was one of the last important mathematicians to hold out against the completed infinite. Since his time mathematics has leaned towards the opposite of his view quoted above: we now take the infinite as real and regard "approaching" and "growing" as figures of speech. This may well be due to the discovery of uncountable sets, which will be discussed in the next section, but it is also due to the declining role of *time* and *motion* in mathematical imagery.

The heyday of the concept of time in mathematics was from 1650 to 1800, when infinitesimal calculus brought almost the whole physical world within the scope of mathematics. Calculus did not just solve problems of change and motion—change and motion were thought to be *fundamental concepts* of calculus. For example, in 1671 Newton believed that all problems about curves reduce to the following two [42, vol. 3, p. 71]:

1. Given the length of space continuously (that is, at every time), to find the speed of motion at any time proposed.

2. Given the speed of motion continuously, to find the length of space described at any time proposed.

Today, the very idea of "change" or "process" is a figure of speech in mathematics because mathematical existence is timeless. For example, the "process" of producing natural numbers by repeatedly adding 1 is *not* something that takes place in time. We are inclined to visualize someone writing numbers on a piece of paper, say, but this is just a mental image. One does not really believe that there is a largest number that has been created so far, and that larger numbers will exist tomorrow. If natural numbers exist, they *all* exist. In the timeless world of mathematical objects, the potential infinite is an actual infinite.

The link between timelessness, actual infinity, and uncountability is forged by the line $\mathbb{R}$ of real numbers that we introduced in Section 2.1. As a geometric object, the line is almost as simple as can be—only points are simpler—but *the relationship between points and the line is almost unfathomable*. In Chapter 1 we saw the complications that result from the existence of irrational numbers, such as $\sqrt{2}$, but these are nothing compared to the problem of grasping *all* irrational numbers. The first to glimpse the depth of this problem were the nineteenth-century German mathematicians Richard Dedekind and Georg Cantor.

Exercises

Two other important examples of countable sets are the set

$$\mathbb{Z} = \{\ldots, -3, -2, -1, 0, 1, 2, 3, \ldots\}$$

of *integers*, and the set $\mathbb{N} \times \mathbb{N}$ of ordered pairs (m, n) of natural numbers.

9.2.1 Explain how to arrange the integers in a list

$$0, -1, 1, \ldots,$$

so that each integer occurs at some finite position.

9.2.2 Explain how to list all pairs (m, n) of positive integers according to the sums $m + n$, starting with $m + n = 2$.

9.2.3 Given a list $r_1, r_2, r_3, \ldots$ of all positive rationals, how can one obtain a list of all rationals?

9.3 The Uncountable

If all points of the straight line fall into two classes such that every point of the first class lies to the left of every point of the second class, then there exists one and only one point which produces this division of all points into two classes, this severing of the straight line into two portions ...

the majority of my readers will be very much disappointed in learning that by this commonplace remark the secret of continuity is to be revealed.

Translation by W. W. Beman of Dedekind's *Continuity and Irrational Numbers*, Open Court 1901, p. 11.

As we saw in Section 1.3, the Pythagoreans discovered that rational numbers do not fill the whole line: there are *gaps* where irrationals occur, such as $\sqrt{2}$ or $\sqrt{3}$. A century or so later, Eudoxus realized that the gaps themselves behave like numbers: a gap has the width of a single point, and it is possible to compare, add, and multiply gaps by reference to their positions among the rationals (Section 1.5). However, the Greeks could deal only with individual gaps, because they did not accept completed infinities. The idea of considering the totality of rationals, let alone the totality of gaps, was taboo, as it still was in the time of Gauss.

It was Dedekind in 1858 who first saw the continuous line as the totality of rationals plus the totality of gaps (or *cuts* in the rationals, as he called them). The line is continuous—without gaps—for the simple reason that all gaps are included in it! This may sound like a joke, but if so it is a very good one. You may prefer to fill each gap by an irrational number, such as $\sqrt{2}$, but there is really no better way to describe an irrational number than by the gap it produces in the rationals.

Dedekind's definition of the line is as simple as possible, but not simpler. It does demand that we accept the infinity of gaps in the rationals as a completed whole. And the infinity of gaps cannot be explained away as merely a potential infinity, the way the infinity of rationals can. Incredible as it may seem, there are *more* gaps than there are rationals, because *the set of gaps is not countable.* Hence accepting the line as a set of numbers means breaking the taboo against actual infinity, and entering the previously unknown world of uncountable sets.

Why ℝ Is an Uncountable Set

Uncountability was discovered by Cantor in 1874, when he showed that the set ℝ of real numbers is uncountable. (It follows that the set of irrationals is also uncountable, and hence that the countable set of rationals has uncountably many gaps.) The proof is by showing that *no countable set of real numbers is all of* ℝ. Cantor's 1874 proof is rather subtle, and I prefer the following, which is based on an idea of the German mathematician Adolf Harnack from 1885.[1]

Harnack shows that a countable set of numbers can be covered by pieces of the line whose total length is small—certainly less than the length of the whole line—hence a countable set does not fill the whole line. Here is a simple way to do it, which shows that covering a countable set is basically the same as finding an infinite series with finite sum.

Suppose that $\{x_1, x_2, x_3, x_4, \ldots\}$ is a countable set of real numbers. Let us cover the point x_1 by a line segment of length 0.1, for example the segment from $x_1 - 0.05$ to $x_1 + 0.05$. Similarly, cover x_2 by a segment of length 0.01, x_3 by a segment of length 0.001, x_4 by a segment of length

---

[1] Actually, Harnack misunderstood the phenomenon he had discovered, and did not realize that it implies uncountability of ℝ. What follows is a correct working out of his idea, with an explanation of what it really implies.

0.0001, and so on. Then the whole countable set $\{x_1, x_2, x_3, x_4, \ldots\}$ is covered by a set of line segments of total length at most

$$0.1 + 0.01 + 0.001 + 0.0001 + \cdots = 0.1111\cdots = 1/9.$$

But the whole line $\mathbb{R}$ has infinite length, so it is not completely covered by these intervals. Therefore, the countable set $\{x_1, x_2, x_3, x_4, \ldots\}$ does not include all real numbers. □

The proof above shows that any countable set can be covered by intervals of total length at most $1/9$, but obviously we did not have to choose the lengths $0.1, 0.01, 0.001, 0.0001, \ldots$ to cover successive points $x_1, x_2, x_3, x_4, \ldots$ in the set. We could have chosen lengths only $1/10$ of these, and got total length at most $1/90$; or lengths only $1/100$ of these, and got total length $1/900$—or whatever. *For any length l, no matter how small, a countable set can be covered by intervals of total length at most l.* Thus, if we consider the countable set itself to have a length, this length can only be zero.

In particular, the set of rational numbers has total length zero, and because of this we say *almost all real numbers are irrational.* In general, we say that "almost all" numbers have a certain property if the set of numbers lacking the property has length zero. It seems remarkable that almost all numbers are irrational when specific irrational numbers, such as $\sqrt{2}$, are hard to find. However, we will see in Section 9.5 that even stranger properties are true for "almost all" numbers. Sometimes the easiest way to show that there are numbers with strange properties is to show that only countably many numbers are *not* strange!

Exercises

Another surprising application of geometric series to the real numbers occurs in the construction of the so-called *Cantor set.* This set consists of the numbers x between 0 and 1 that remain after the following series of removals.

First remove the *middle third,* consisting of all x such that $1/3 < x < 2/3$. Next remove the middle thirds of the two intervals that remain; namely, remove the x such that $1/9 < x < 2/9$ and $7/9 < x < 8/9$. In general, at stage $n + 1$ remove the middle thirds of all intervals that remain after stage n. Figure 9.1 shows the results of the first six stages.

Figure 9.1: Stages in the construction of the Cantor set.

It is important that when we remove a middle third we do *not* remove its endpoints: for example, when we remove the first middle third we do not remove the points 1/3 and 2/3. Because of this, there are many points in the Cantor set, such as 0, 1, 1/3, 2/3, 1/9, 2/9, 7/9, 8/9, ...

9.3.1 Show that the length removed at stage 1 is 1/3; at stage 2 it is 2/9; at stage 3 it is 4/27.

9.3.2 Conclude that the total length of the intervals removed is the infinite series

$$\frac{1}{3} + \frac{2}{9} + \frac{4}{27} + \cdots$$

9.3.3 The series just found is a geometric series $a + ar + ar^2 + \cdots$. What are the values of a and r?

9.3.4 Conclude the total length removed, $\frac{a}{1-r}$, equals 1.

Thus the Cantor set has total length zero! This is not completely surprising, since we have just seen that infinite sets can have total length zero—if they are countable. However, we will see in the next exercise set that the Cantor set is *not* countable.

9.4 The Diagonal Argument

The Harnack proof that countable sets have length zero shows in dazzling style that any countable set $\{x_1, x_2, x_3, x_4, \ldots\}$ fails to be all of $\mathbb{R}$. But what if someone (say, the legendary guy from Missouri) asks to be *shown* a number x not on the infinite list $x_1, x_2, x_3, x_4, \ldots$? The proof also tells where to find such an x, if you look closely. This is definitely worth the trouble, even if you delight in the Harnack proof, because it reveals a more direct path to uncountability.

We again cover $x_1, x_2, x_3, x_4, \ldots$ by intervals, but now we also construct x, between 0 and 1 say, so that x is

> outside the interval containing x_1,
>
> and outside the interval containing x_2,
>
> and outside the interval containing x_3,
>
> and outside the interval containing x_4,
>
> and outside the interval containing x_5,
>
> and so on.

This is easy because:

1. We can take the interval containing x_1 to consist of all numbers agreeing with x_1 up to and including the first decimal place. For example, if $x_1 = 3.14159\cdots$ take the interval from 3.1 to 3.2. Then an x *outside* this interval is any number that disagrees with x_1 in the first decimal place.

2. We can take the interval containing x_2 to consist of all numbers agreeing with x_2 up to and including the second decimal place. Then an x *outside* this interval is any number that disagrees with x_2 in the second decimal place.

3. Similarly, we can take intervals so that x disagrees with x_3 in the third decimal place, with x_4 in the fourth decimal place, with x_5 in the fifth decimal place, and so on.

Thus, if the intervals covering $x_1, x_2, x_3, x_4, \ldots$ are chosen as above, a real number x outside all these intervals can be constructed simply by *avoiding* the first decimal place of x_1, then the second decimal place of x_2, the third decimal place of x_3, the fourth decimal place of x_4, and so on. At this point one realizes that the intervals are actually irrelevant: *we can make x different from each of $x_1, x_2, x_3, x_4, \ldots$ by making x disagree with each x_n in the nth decimal place.*

This argument for showing that a countable set does not include all real numbers was devised by Cantor in 1891 (though it is perhaps implicit in the earlier proofs that $\mathbb{R}$ is uncountable). It is often called the *diagonal argument* or *diagonalization* because it uses only the "diagonal" digits in the list of decimal expansions of $x_1, x_2, x_3, x_4, \ldots$ to construct a new number x. For example, if

$$x_1 = 3.\underline{1}4159\cdots$$
$$x_2 = 2.1\underline{7}281\cdots$$
$$x_3 = 0.54\underline{7}71\cdots$$
$$x_4 = 1.414\underline{2}1\cdots$$
$$x_5 = 1.7322\underline{1}\cdots$$

$$\vdots \qquad\qquad\qquad\qquad (1)$$

then we make x disagree with each of the underlined digits in turn: thus x_1 does *not* have 1 in the first decimal place, nor 7 in the second, and so on.

The only danger in the argument is the possibility of producing an x which, while it has different digits from each x_n, is nevertheless equal to one of them—the way $0.999\cdots$ equals $1.000\cdots$ for example. Equal numbers with different digits involve the strings $000\cdots$ or $999\cdots$, so we can avoid the danger by not using 0 or 9 in x. Specifically, we can compute x by the rule

$$n\text{th decimal place of } x = \begin{cases} 2 & \text{if } n\text{th place of } x_n \text{ is 1,} \\ 1 & \text{otherwise,} \end{cases}$$

which gives $x = 0.21112\cdots$ for the list (1) above. The x produced by this rule has different digits from all of $x_1, x_2, x_3, x_4, x_5, \ldots$, so x is necessarily *unequal* to $x_1, x_2, x_3, x_4, x_5, \ldots$, because x has no 0 or 9 after the decimal point.

Exercises

When we applied the diagonal argument to decimal expansions of real numbers, we had to take a little care in choosing new digits in order to avoid numbers with two different decimal expansions. This difficulty does not occur with objects that can be written in only one way.

Our first example consists of the sets S of natural numbers. Each S can be described by an infinite sequence of 0s and 1s, with 1 in the nth place if and only if $n \in S$.

9.4.1 Which set is described by the sequence 01010101...?

9.4.2 Write down the first 16 digits of the sequence that describes the set of squares.

9.4.3 Given sequences that describe sets $S_1, S_2, S_3, \ldots$ of natural numbers, by what rule can we define a set S that differs from each S_n with respect to the number n?

It follows that the list $S_1, S_2, S_3, \ldots$ does not include all sets of natural numbers. In fact, it does not include the set S defined by

$$n \in S \Leftrightarrow n \notin S_n.$$

Our second example—which is just a variation of the first—consists of all *infinite paths* in the *binary tree* shown in Figure 9.2. Obviously,

Figure 9.2: The infinite binary tree.

any path down this tree can be described by a sequence of 0s and 1s: 0 when the path goes left at a vertex, 1 when it goes right. Thus there are uncountably many infinite paths down the tree.

9.4.4 Show, on the other hand, that there are only countably many vertices in the tree.

The infinite paths are interesting because they correspond to points in the Cantor set.

9.4.5 By comparing Figure 9.2 with Figure 9.1, show that each infinite path in the tree corresponds to a point common to all the sets shown in black in Figure 9.1; that is, to a point *not* removed, and hence to a point in the Cantor set.

9.4.6 Conclude that the Cantor set is uncountable.

9.5 The Transcendental

In 1874, Cantor was well aware that the world was not ready for un-countable sets. His revolutionary discovery is hidden in a paper whose title (translated from German) is *On a property of the collection of all real algebraic numbers*, and the property he demonstrates is in fact the *countability* of this collection. Uncountability slips in quietly as he pro-ceeds to show that, for any countable set $\{x_1, x_2, x_3, x_4, x_5, \ldots\}$, we can find a real number $x \neq x_1, x_2, x_3, x_4, x_5, \ldots$. When $\{x_1, x_2, x_3, x_4, x_5, \ldots\}$ is taken to be the set of algebraic numbers, his construction therefore gives an x that is *not algebraic*. This was interesting, because at the time very few examples of nonalgebraic numbers were known.

Examples of algebraic numbers are the rational numbers and cer-tain irrational numbers such as $\sqrt{2}$. The definition is that x is *algebraic* if it satisfies a polynomial equation with integer coefficients, that is, an equation of the form

$$a_n x^n + a_{n-1} x^{n-1} + \cdots + a_1 x + a_0 = 0, \quad \text{for integers } a_n, a_{n-1} \ldots, a_1, a_0.$$

Thus $\sqrt{2}$ is algebraic because it satisfies the equation $x^2 - 2 = 0$. Simi-larly, $\sqrt[3]{5}$ is algebraic because it satisfies $x^3 - 5 = 0$. It is also true, though not so obvious, that $\sqrt{2} + \sqrt[3]{5}$ is algebraic. In fact, so is any number "ex-pressible by radicals," that is, built from integers by a finite number of applications of the operations $+, -, \times, \div$, and radicals (square roots, cube roots, and so on). Even this does not exhaust all the algebraic numbers, because the roots of $x^5 + x + 1 = 0$ are algebraic numbers *not* expressible by radicals.

Thus it is not easy to write down a list of all algebraic numbers, and not very rewarding either—unless you want an easy proof that nonal-gebraic numbers exist. Before 1874, all known proofs of this fact relied on some hard work on the structure of algebraic numbers. Cantor's in-sight was that algebra can be completely avoided if we put the work into reasoning about infinity instead; namely, in proving that *the alge-braic numbers form a countable set*.

Following an idea of Dedekind, Cantor did this as follows. To each polynomial with integer coefficients,

$$p(x) = a_n x^n + a_{n-1} x^{n-1} + \cdots + a_1 x + a_0,$$

he assigned a natural number called its "height:"

$$\text{height}(p) = n + |a_n| + |a_{n-1}| + \cdots + |a_1| + |a_0|.$$

The only important thing about height is that *there are only finitely many polynomials with a given height.* This is because the degree n of the polynomial p must be $\leq \text{height}(p)$, and each of the $n+1$ coefficients of p must have absolute value $\leq \text{height}(p)$.

So in principle we could write down a finite list (in alphabetical order, say) of all the polynomials with a given height. Then, if we string together

> the list of polynomials of height 1,
>
> the list of polynomials of height 2,
>
> the list of polynomials of height 3,
>
> and so on,

we get a list of all polynomials. It is an infinite list, but each polynomial on it has only finitely many predecessors. Finally, if we replace each polynomial on the list by the finite list of its solutions, we get a list of all algebraic numbers. Thus the set of algebraic numbers is countable, and similarly (by listing only real solutions) the set of real algebraic numbers is also countable. □

Cantor then applies his method for finding a real number x different from all members of a countable set, so the result is a nonalgebraic x. Nonalgebraic numbers are also called *transcendental,* because they "transcend" definition by ordinary (algebraic) means. Such numbers were discovered by the French mathematician Joseph Liouville in 1844, who showed that numbers such as

$$0.101001000000100000000000000000000000001000\ldots$$

are transcendental when the blocks of zeros grow sufficiently rapidly in length (in this case the lengths of the blocks are 1, 2×1, $3 \times 2 \times 1$, $4 \times 3 \times 2 \times 1, \ldots$).

The first to prove a "naturally occurring" number to be transcendental was Liouville's compatriot Charles Hermite in 1873. Hermite made sophisticated use of calculus to prove that the number e (mentioned in Section 2.7) cannot satisfy a polynomial equation with integer coefficients. In 1882 his method was extended by the German mathematician Ferdinand Lindemann to prove that π is transcendental. These hard-won results gave the impression that transcendental numbers are rare, but in fact the opposite is the case. Since there are only countably many algebraic numbers, it follows by Harnack's theorem that the set of real algebraic numbers has zero length, and hence *almost all real numbers are transcendental.*

Squaring the Circle

> Well I now apply the straight rod—so—thus squaring the circle: and there you are.
>
> > Spoken by the character Meton in Aristophanes' *The Birds*, Act 1.
>
> **To square the circle.** To attempt an impossibility. The allusion is to the impossibility of exactly determining the precise ratio (π) between the diameter and the circumference of a circle, and thus constructing a circle of the same area as a given square.
>
> *Brewer Dictionary of Phrase and Fable*, HarperResource 2000.

Aristophanes wrote *The Birds* around 400 BCE, so apparently "squaring the circle" has been synonymous with futility and folly for more than 2,000 years. We mentioned the problem briefly in Sections 4.4 and 4.9, but what is it precisely, and how is it related to the transcendence of π?

First, the *Brewer Dictionary* is a little inaccurate. The problem is really to construct a square equal in area to a given circle and, more importantly, the "construction" has to be by ruler and compass. This is the kind of construction you probably learned in high school geometry; it involves

- drawing the line through two given points,

- drawing the circle with a given center and through a given point,

- finding the intersections of the lines and circles drawn, and using them to draw new lines and circles.

If the construction begins with two points a unit distance apart (say, at the center and on the circumference of the circle we wish to "square"), then coordinate geometry shows that all constructible lengths result from 1 by a finite number of applications of the operations +, −, ×, ÷, and square roots (see the exercises below). Thus *all constructible numbers are algebraic.* Lindemann's 1882 proof therefore shows that π *is not a constructible number,* and it follows that the circle cannot be squared according to the rules laid down by the Greeks.

It also shows that Descartes was not completely off target when he asserted that the lengths of curves "cannot be known by human minds" (see the beginning of Chapter 4). Descartes accepted only algebraic construction processes, such as forming intersections of circles and parabolas. These go beyond the Greek rules, but they yield only algebraic numbers, so Lindemann's result is still relevant: the length π of the unit semicircle cannot be *algebraically* known.

However, as we saw in Chapter 4, π can be known through the infinite series for $\pi/4$:

$$1 - \frac{1}{3} + \frac{1}{5} - \frac{1}{7} + \frac{1}{9} - \frac{1}{11} + \cdots .$$

Since π cannot be known algebraically, this formula for π is a fine example of *knowledge that only infinity can give us.* When infinity gives such gifts, who can doubt that it exists?

Exercises

The ruler and the compass allow us to construct straight lines, which typically have equations of the form $y = mx + c$ (as we saw in Section 3.2), and circles, which have equations of the form

$$(x - a)^2 + (y - b)^2 = r^2,$$

where (a, b) is the center of the circle and r is its radius. Moreover, the constants m, c, a, b, r in these equations can be calculated from points defining the line or circle by the operations $+, -, \times, \div$ of arithmetic. For example:

9.5.1 Show that if the line $y = mx + c$ goes through the points (a_1, b_1) and (a_2, b_2) then

$$m = \frac{b_2 - b_1}{a_2 - a_1} \quad \text{and} \quad c = \frac{a_2 b_1 - b_2 a_1}{a_2 - a_1}.$$

All further constructible points arise from intersections of lines and circles, from which we construct new lines and circles, and so on. The coordinates of these points result from the constants in the equations of the lines and circles by $+, -, \times, \div$ and the $\sqrt{}$ operation, which is why they are algebraic. Only these operations are needed, because all the equations that need to be solved are either linear or quadratic.

9.5.2 Show that the lines $y = m_1 x + c_1$ and $y = m_2 x + c_2$ intersect at the point where

$$x = \frac{c_2 - c_1}{m_1 - m_2}, \quad y = \frac{m_1(c_2 - c_1)}{m_1 - m_2} + c_1.$$

9.5.3 Show that the line $y = mx + c$ and the circle $(x - a)^2 + (y - b)^2 = r^2$ meet at points found from m, c, a, b, r by solving a quadratic equation.

9.5.4 Show that the intersection of the circles

$$(x - a_1)^2 + (y - b_1)^2 = r_1^2 \quad \text{and} \quad (x - a_2)^2 + (y - b_2)^2 = r_2^2,$$

that is

$$x^2 - 2a_1 x + a_1^2 + a_1^2 + y^2 - 2b_1 y + b_1^2 = r_1^2 \quad \text{and}$$
$$x^2 - 2a_2 x + a_2^2 + a_2^2 + y^2 - 2b_2 y + b_2^2 = r_2^2,$$

also lies on

$$2(a_1 - a_2)x + 2(b_1 - b_2)y + a_2^2 - a_1^2 + b_2^2 - b_1^2 = r_2^2 - r_1^2,$$

and that the latter is the equation of a line.

9.5.5 Deduce from Exercises 9.5.4, 9.5.3, and 9.5.2 that all constructible points may be found by solving linear and quadratic equations, and hence that their coordinates are algebraic.

9.6 Yearning for Completeness

The last section of a book is where one would like to tie up all loose ends, to achieve completeness and closure. The real numbers $\mathbb{R}$ are the ideal topic for this purpose. They end the search for numbers such as $\sqrt{2}$ and π, they close all gaps in the rational numbers, and (with the help of $\mathbb{R}^2, \mathbb{R}^3, \mathbb{R}^4, \ldots$) they form a sound foundation for geometry—in flat or curved spaces, and in any number of dimensions—and the algebras of complex numbers and quaternions. Also, the idea of an infinite *set*, so crucial for bringing the real numbers into existence, is essential for the concepts of *ideal* and *equivalence class*, now so important in number theory, geometry, topology, and (perhaps) astronomy.

The real numbers are equally important as a basis for the limit concept in calculus, the key to the modern approach which avoids the contradictory aspects of infinitesimals. If one has an increasing sequence of numbers, such as

$$\frac{1}{2}, \quad \frac{3}{4}, \quad \frac{7}{8}, \quad \frac{15}{16}, \quad \frac{31}{32}, \quad \ldots,$$

and if all members of the sequence are below some bound (in this case 1, or any larger number), then one expects the sequence to have a *limit*: a least number greater than each member of the sequence. In this example the limit is the rational number 1. But of course we cannot always expect the limit to be rational. For example, the limit of the increasing sequence (bounded above by 1)

$$1 - \frac{1}{3}, \quad 1 - \frac{1}{3} + \frac{1}{5} - \frac{1}{7}, \quad 1 - \frac{1}{3} + \frac{1}{5} - \frac{1}{7} + \frac{1}{9} - \frac{1}{11}, \quad \ldots$$

is $\pi/4$. If a sequence approaches some gap in the rational numbers from below, then its limit is guaranteed to exist only if all gaps in the rationals are filled by real numbers. This is why the set of *all* real numbers is needed for calculus: we want all bounded increasing sequences to have limits.

Thus, with hindsight, it begins to look as though most of the struggles with the impossible in the history of mathematics have been part of the struggle to extend the number concept. This may be so, but it does not mean that we have reached complete understanding of the real numbers, even of individual numbers as simple as $\sqrt{2}$. Back in

Section 1.5 we mentioned that very little is known about the infinite decimal for $\sqrt{2}$, which begins with

$$1.4142135623730950488016887242209698078569\cdots.$$

It seems quite likely that each of the 10 digits 0, 1, 2, 3, 4, 5, 6, 7, 8, 9 occurs 1/10 of the time, on the average—why would one digit occur more frequently than any other?—but nothing whatever in this direction has been proved. In fact this equal-frequency property, called *normality*, has not been proved for any irrational algebraic number, or for any of the "naturally occurring" transcendental numbers such as e or π. *Yet almost all real numbers are normal!* This follows from an elaboration of the common sense idea that, in a random infinite decimal, none of the digits 0, 1, 2, 3, 4, 5, 6, 7, 8, 9 should occur more frequently than another.

Another thing we do not understand is the so-called *continuum problem*: how big is the infinity of real numbers? In particular, is $\mathbb{R}$ the "smallest" uncountable infinity? We know that the natural numbers represent the smallest possible infinity, in the sense that they can be ordered so that each number has only finitely many predecessors. This is impossible for $\mathbb{R}$, since $\mathbb{R}$ is uncountable. The smallest uncountable set has an ordering in which each member has only *countably* many predecessors. Cantor believed that such an ordering exists for $\mathbb{R}$ but was unable to prove it. He ended his life in a mental hospital, and some people believe that it was the continuum problem that put him there.

Some of the deepest and most intricate ideas of twentieth-century mathematics were brought to bear on the continuum problem, without a clearcut result. What we now know is that the problem *cannot be settled* by the accepted axioms of set theory. To some, this is a sign that the problem is impossible to solve; to others, it means only that we are missing something, and that the continuum problem is not as impossible as it looks.

You can probably guess what side I'm on. But even if the continuum problem is solved one day, infinitely many problems about the real numbers will remain open. Our inability to *list* all real numbers implies an inability to *know all facts* about real numbers as well. (In fact, it implies inability to know all facts about natural numbers too, but I don't wish to pursue this thread here—fascinating though it is.)

Thus new ideas, and perhaps new struggles with the apparently impossible, will always be called for—even in the theory of numbers.

Mathematics is the most permanent world, but it is also a nev-erending story.

Epilogue

> From the infinitesimal calculus to the present, it seems to me, the essential progress in mathematics has resulted from successively annexing notions which, for the Greeks or the Renaissance geometers or the predecessors of Riemann, went "outside mathematics" because it was impossible to define them.
>
> Jacques Hadamard, 1905 letter to Borel, translated by G. H. Moore [38, p. 318]

It should now be apparent that "yearning for the impossible" is the source of many advances in mathematics, and we can discern at least two kinds of "impossibility" that have been fruitful—actual and apparent. This is no surprise, in view of the remark of Kolmogorov quoted in the preface, that mathematical discoveries occur in a thin layer between the trivial and the impossible. When the layer is thin, it is hard to tell the difference between actual and apparent impossibility, and both can be close to a truth.

Actual Impossibilities that Nevertheless Lead to New Truths

Sometimes what mathematicians yearn for is too simple to be true, such as a world in which all numbers are rational. But it may approximate a more complicated truth, such as the world of real numbers obtained by filling the gaps in the rationals.

Indeed, some truths may be too complex to reach on the first attempt. We may need to begin with an "approximate truth" that is false, but on the right track, such as the theory of infinitesimals as an approximation to calculus.

Finally, some impossibilities are not even on the right track, but they fall close to a new truth by dumb luck. It is hopeless to yearn for three-dimensional numbers, but in searching for them we bump into the four-dimensional quaternions, which are the nearest thing to numbers in any dimension greater than 2.

Apparent Impossibilities that *Are* New Truths

Most of the examples in this book are in this category: irrational numbers, imaginary numbers, points at infinity, curved space, ideals, and various types of infinity. These ideas seem impossible at first because our intuition cannot grasp them, but they can be captured with the help of mathematical symbolism, which is a kind of technological extension of our senses.

For example, we cannot *see* any difference between the irrational point $\sqrt{2}$ and a rational point close to it, such as 1.41421356. But with the symbolism of infinite decimals we can understand why $\sqrt{2}$ is different from any rational number: its decimal expansion is infinite and not periodic.

The situation is similar, but not quite the same, with $\sqrt{-1}$. It is impossible for $\sqrt{-1}$ to be a real number, since its square is negative. This implies that $\sqrt{-1}$ is neither greater nor less than zero, so we cannot see $\sqrt{-1}$ on the real line. However, $\sqrt{-1}$ behaves like a number with respect to + and ×. This prompts us to *look elsewhere* for it, and indeed we find it on another line (the imaginary axis) perpendicular to the real line.

Impossibility and Mathematical Existence

Until about 100 years ago the concept of impossibility was vague. We have seen, time and time again, how apparently impossible structures were found to be actual. Yet some things certainly *are* impossible, for example, contradictions. An actually existing object cannot have contradictory properties, such as being both square and round. But is contradiction the *only* reason for impossibility?

In a famous address to the International Congress of Mathematicians in 1900, Hilbert stated the "easy direction" of this situation:

> If contradictory attributes be assigned to a concept, I say,
> that *mathematically the concept does not exist.* So, for ex-
> ample, a real number whose square is −1 does not exist
> mathematically.

Then he boldly claimed the converse:

> But if it can be proved that the attributes assigned to the
> concept can never lead to a contradiction by the applica-
> tion of a finite number of logical processes, I say that the
> mathematical existence of the concept (for example, of a
> number or a function which satisfies certain conditions) is
> thereby proved.

It is not clear how Hilbert justified this claim, but it can be justified
by later results obtained by mathematical logicians—Leopold Löwen-
heim in 1915, Thoralf Skolem in 1922, and Kurt Gödel in 1929. Their
results show that consistency implies existence in the following sense:
any consistent set of sentences in first-order logic (a language ade-
quate for mathematics as it is normally understood) has a *model*, that
is, an interpretation that makes all the given sentences true.

The "interpretation" is symbolic, constructed from symbols of the
language in which the sentences are expressed, and hence is not as in-
tuitive as one might hope. However, this is how mathematics has often
absorbed the "impossible" in the past. First, the impossible object is
represented as a symbol, such as $\sqrt{-1}$, then one tests how this symbol
interacts with its fellows from the world of accepted mathematical objects. If the new symbol is compatible, as $\sqrt{-1}$ was found to be, then
it becomes accepted as well, and is reckoned to denote a new mathematical object.

The Future of the Impossible

I would not like to leave the impression that all conflicts in mathematics have now been resolved, and that there is no longer any need to
wrestle with impossibilities. Quite the opposite. For example, physics
has been plagued for the last 80 years by a mathematical conflict as se-
vere as the conflict over infinitesimals in calculus. Its two most important theories, general relativity and quantum theory, are incompatible!

In practice, relativity theory and quantum theory stay out of each other's way by working in different domains: relativity in the big world of astronomy and quantum theory in the small world of the atom. In their respective worlds, these theories have been shown to be amazingly accurate. But there is only one world, so it is impossible for both relativity theory and quantum theory to be correct. Presumably, the truth lies somewhere very close to both of them, but no one has yet been able to express this truth satisfactorily.

There are theories that purport to reconcile relativity and quantum theory, in particular the so-called *string theory*. But, so far, string theory cannot be tested experimentally, so it is not really physics. The surprising thing is that string theory is wonderful mathematics! In the 1990s, string theory was used to solve deep problems in pure mathematics (see Brian Greene's book *The Elegant Universe*), some of them so improbable they are called *moonshine*. If string theory can do this, what is in store when the impossible world of relativity and quantum theory is *really* understood?

References

1. Leon Battista Alberti, *On Painting*.
   Translated by John R. Spencer.
   New Haven, CT: Yale University Press, 1966.

2. Aristophanes, *The Birds and Other Plays*.
   Translated by David Barrett and Alan Sommerstein.
   London, UK: Penguin Classics, 2003.

3. Benno Artmann, *Euclid—The Creation of Mathematics*.
   New York, NY: Springer-Verlag, 1999.

4. George Berkeley, *De Motu and the Analyst*.
   Edited by Douglas Jesseph.
   Dordrecht: Kluwer Academic Publishers, 1992.

5. Bill Bryson, *In a Sunburned Country*.
   New York, NY: Broadway, 2001.

6. Girolamo Cardano, *Ars Magna or the Rules of Algebra*.
   Translated by T. Richard Witmer.
   New York, NY: Dover, 1993.

7. Lewis Carroll, *Alice's Adventures in Wonderland*.
   New York, NY: Signet Classics, 2000.

8. Sir Arthur Conan Doyle, *The Sign of Four*.
   London, UK: Penguin Classics, 2001.

9. H. S. M. Coxeter, *Regular Polytopes*.
   New York, NY: Dover, 1973.

10. Dante Alighieri, *The Divine Comedy*.
    Translated by Mark Musa.
    London, UK: Penguin Classics, 2003.

11. Philip J. Davis, *The Mathematics of Matrices*.
    Waltham, MA: Blaisdell, 1965.

12. Richard Dedekind, *Essays on the Theory of Numbers.*
    I. Continuity and Irrational Numbers.
    Translated by W. W. Beman.
    Chicago, IL: Open Court, 1901.

13. Richard Dedekind, *Theory of Algebraic Integers.*
    Translated by John Stillwell.
    Cambridge, UK: Cambridge University Press, 1996.

14. René Descartes, *The Geometry of René Descartes.*
    Translated by David E. Smith and Marcia L. Latham.
    New York, NY: Dover, 1954.

15. M. C. Escher, *Escher on Escher. Exploring the Infinite.*
    New York, NY: Harry N. Abrams, 1989.

16. Euclid, *The Thirteen Books of the Elements.*
    Edited by Sir Thomas Heath.
    New York, NY: Dover, 1956.

17. Leonhard Euler, *Elements of Algebra.*
    Translated by John Hewlett.
    New York, NY: Springer-Verlag, 1984.

18. Leonhard Euler, *Introduction to the Analysis of the Infinite.*
    Translated by John D. Blanton.
    New York, NY: Springer-Verlag, 1988.

19. Fibonacci, *The Book of Squares.*
    Translated by L. E. Sigler.
    Boston, MA: Academic Press, 1987.

20. Robert Graves, *Complete Poems*, Volume 3.
    Manchester, UK: Carcanet Press, 1999.

21. R. P. Graves, *The Life of Sir William Rowan Hamilton.*
    New York, NY: Arno Press, 1975.

22. Brian Greene, *The Elegant Universe.*
    New York, NY: W. W. Norton, 1999.

23. J. Hadamard, *The Psychology of Invention in the Mathematical Field.*
    New York, NY: Dover, 1954

24. W. R. Hamilton, *The Mathematical Papers of Sir William Rowan Hamilton.*
Edited by A. W. Conway and J. L. Synge.
Cambridge, UK: Cambridge University Press, 1931.

25. Thomas L. Heath, *Diophantus of Alexandria.*
New York, NY: Dover, 1964.

26. David Hilbert, *Foundations of Geometry.*
Translated by Leo Unger.
La Salle, IL: Open Court, 1971.

27. David Hilbert and Stefan Cohn-Vossen, *Geometry and the Imagination.*
Translated by P. Nemenyi.
New York, NY: Chelsea Publishing Co., 1952.

28. Thomas Hobbes, *The English Works of Thomas Hobbes*, Volume 7.
Edited by Sir William Molesworth.
London, UK: J. Bohn, 1839–1845.

29. Oliver Wendell Holmes, *The Autocrat of the Breakfast Table.*
Pleasantville, NY: Akadine Press, 2002.

30. Douglas Jesseph, *Squaring the Circle.*
Chicago, IL: University of Chicago Press, 2000.

31. A. N. Kolmogorov, *Kolmogorov in Perspective.*
Providence, RI: American Mathematical Society, 2000.

32. Adrien-Marie Legendre, *Théorie des Nombres.*
Paris: Courcier, 1808.

33. Marquis de l'Hopital, *Analyse des infiniments petits.*
Paris: Imprimerie Royale, 1696.

34. Elisha Scott Loomis, *The Pythagorean Proposition.*
Washington, DC: National Council of Teachers of Mathematics, 1968.

35. Lucretius, *The Way Things Are.*
Translation of *De rerum natura* by Rolfe Humphries.
Bloomington, IN: Indiana University Press, 1968.

36. D. MacHale, *Comic Sections.*
Dublin, Ireland: Boole Press, 1993.

37. Barry Mazur, *Imagining Numbers.*
New York, NY: Farrar Straus Giroux, 2002.

38. G. H. Moore, *Zermelo's Axiom of Choice.*
New York, NY: Springer-Verlag, 1982.

39. Paul Nahin, *An Imaginary Tale*.
    Princeton, NJ: Princeton University Press, 1998.

40. Joseph Needham, *Science and Civilisation in China*.
    Cambridge, UK: Cambridge University Press, 1954.

41. Tristan Needham, *Visual Complex Analysis*.
    Oxford, UK: Oxford University Press, 1997.

42. Isaac Newton, *The Mathematical Papers of Isaac Newton*.
    Cambridge, UK: Cambridge University Press, 1967–1981.

43. Nicomachus, *Introduction to Arithmetic*.
    New York, NY: Macmillan, 1926.

44. Jean Pèlerin, *De artificiali perspectiva*.
    Toul, France: Petri Iacobi, 1505.

45. John Ratcliffe, *Foundations of Hyperbolic Manifolds*, Second Edition
    New York, NY: Springer, 2006.

46. Adrian Room, ed., *Brewer's Dictionary of Phrase and Fable*.
    New York, NY: HarperResource, 2000.

47. Girolamo Saccheri, *Girolamo Saccheri's Euclides vindicatus*.
    Translated by G. B. Halsted.
    Chicago, IL: Open Court, 1920.

48. Michael Spivak, *A Comprehensive Introduction to Differential Geometry*.
    Berkeley, CA: Publish or Perish, 1979.

49. Simon Stevin, *The Principal Works of Simon Stevin*.
    Amsterdam: C.V. Swets and Zeitlinger, 1955–1964.

50. John Stillwell, *Mathematics and Its History*, Second Edition.
    New York, NY: Springer-Verlag, 2002.

51. John Stillwell, *Sources of Hyperbolic Geometry*.
    Providence, RI: American Mathematical Society, 1996.

52. Jeffrey Weeks, *The Shape of Space*, Second Edition.
    New York, NY: Marcel Dekker, 2001.

Index